Impact of Pollution on Animal Products

NATO Science for Peace and Security Series

This Series presents the results of scientific meetings supported under the NATO Programme: Science for Peace and Security (SPS).

The NATO SPS Programme supports meetings in the following Key Priority areas: (1) Defence Against Terrorism; (2) Countering other Threats to Security and (3) NATO, Partner and Mediterranean Dialogue Country Priorities. The types of meeting supported are generally "Advanced Study Institutes" and "Advanced Research Workshops". The NATO SPS Series collects together the results of these meetings. The meetings are co-organized by scientists from NATO countries and scientists from NATO's "Partner" or "Mediterranean Dialogue" countries. The observations and recommendations made at the meetings, as well as the contents of the volumes in the Series, reflect those of participants and contributors only; they should not necessarily be regarded as reflecting NATO views or policy.

Advanced Study Institutes (ASI) are high-level tutorial courses intended to convey the latest developments in a subject to an advanced-level audience

Advanced Research Workshops (ARW) are expert meetings where an intense but informal exchange of views at the frontiers of a subject aims at identifying directions for future action

Following a transformation of the programme in 2006 the Series has been re-named and re-organised. Recent volumes on topics not related to security, which result from meetings supported under the programme earlier, may be found in the NATO Science Series.

The Series is published by IOS Press, Amsterdam, and Springer, Dordrecht, in conjunction with the NATO Public Diplomacy Division.

Sub-Series

A.	Chemistry and Biology	Springer
B.	Physics and Biophysics	Springer
C.	Environmental Security	Springer
D.	Information and Communication Security	IOS Press
E.	Human and Societal Dynamics	IOS Press

http://www.nato.int/science
http://www.springer.com
http://www.iospress.nl

Series C: Environmental Security

Impact of Pollution on Animal Products

edited by

Bernard Faye

CIRAD-Centre de Coopération Internationale en
Recherche Agronomique pour le Développement,
Montpellier, France

and

Yuriy Sinyavskiy

KAN - KAZAKH Academy of Nutrition,
Almaty, Kazakhstan

 Springer

Published in cooperation with NATO Public Diplomacy Division

Proceedings of the NATO Advanced Research Workshop on
Impact of Pollution on Animal Products
Almaty, Kazakhstan
27–30 September, 2007

A C.I.P. Catalogue record for this book is available from the Library of Congress.

ISBN 978-1-4020-8358-7 (PB)
ISBN 978-1-4020-8357-0 (HB)
ISBN 978-1-4020-8359-4 (e-book)

Published by Springer,
P.O. Box 17, 3300 AA Dordrecht, The Netherlands.

www.springer.com

Printed on acid-free paper

CONTENTS

Foreword

The international advanced research workshop funded by NATO and entitled "impact of pollutions on animal and animal products" was organized at Almaty (Kazakhstan) on 27–30 September 2007. Thirty-one scientists from 12 countries (Kazakhstan, Kirgizstan, Azerbaijan, Ukraine, Russia, France, Great Britain, Italy, Belgium, Romania and Morocco) presented conferences at this meeting to share their experience and results. The programme included three main aspects: (i) generality on the pollution situation in Central Asia and former Soviet Union republics, (ii) the pollution area and pollution origin in Central Asia and Western countries in relation with animal health, and (iii) the relationships between soil contamination, plant contamination and animal products status. The present workshop contributed highly to the exchange between scientists giving the opportunity for researchers from Central Asia to access to new scientific approaches and methodologies, and for European scientists to assess the extent of the environmental problems in this part of the world. No doubt that these exchanges were the main success of the workshop marked by very stimulating discussions. Such meeting was also the opportunity to put on the first stone of a scientific network focused on the subject of the workshop.

The importance of pollution in Central Asia in general and in Kazakhstan in particular is a well-known feature and several references are available on the source and localization of pollution problems in those countries. The references are also abundant on the impact of the environmental failures on human health. The Kazakh Academy of Nutrition, one of the partners of this workshop is involved for several years in food safety assessment for human consumers. On contrary, the place of the animals between contaminations of the resources for animal feeding and drinking, and the human consumers of animal products is widely understudied.

Yet, animals play several roles in the links between environment and human health. The domestic and wild animals, both terrestrial and aquatic ones, could be considered under three aspects: (i) they are the final receptor of pollutants and therefore, they can be affected by the chronic or acute toxicity of organic and non-organic pollutants presents in their environment or in their food; (ii) they can be a "sentinel" of the polluting status of the environment and therefore, the assessment of their status vis-à-vis the pollutants is quite essential to identify the environmental risks; and (iii) they are a provider of pollutants for the human consumers through their products (milk, meat, eggs, wastes) with sometimes a role in the chronicle concentration of some undesirable molecules in food.

For all theses aspects, several research fields have to be solicited: veterinary sciences, epidemiology, ecology, chemistry, biochemistry, animal and human nutrition, ecotoxicology, animal production and so on. So, the study of the impact of pollution on animals and animal products needs absolutely a multidisciplinary approach including holistic and analytical aspects. Thus, in the perspectives for a future collaboration between the participants of this workshop, it is expected to propose a concept note for a common project including several disciplines and field of research in order to cover the complexity of the role of livestock in the pollutant flow between environment and man.

The final recommendations of the workshop were proposed in that way, with the clear objective to construct a common project by identifying the priorities in further researches and by proposing the strategy for prevention of pollutants in animal products in relation with policy makers. Such concept note has to be proposed in the frame of the

7th UE community research programme or other funding agencies having the environment preservation and protection as finality.

Finally, the co-directors of the advanced research workshop are indebted to NATO and secondly to the French Embassy in Kazakhstan for their support in this meeting. Three institutions were involved in the general organization: the international cooperation centre for agronomic research in development (CIRAD-France), the Kazakh Academy of Nutrition at Almaty and the Al-Farabi University (KazGu, Kazakhstan), especially the department of bio-technology, biochemistry and plant technology. Their collaboration was able to overcome the traditional difficulties for organizing international meeting in spite of the language barrier and of the different scientific culture. It must be recalled also that this meeting was possible because an already long scientific collaboration was existing between the partners of the organizing committee. All of them consider the Advanced Research Workshop organized at Almaty as an essential milestone of their current cooperation on milk quality.

Dr Bernard FAYE and Pr Yuriy SINYAVSKIY
Co-directors and scientific editors

OPENING SESSION

Generalities on the role of institutions in the field of pollution assessment for animal and human health

Abstracts of the opening ceremony and full conference of L. Astanina

HEALTH OF MAN: YESTERDAY, TODAY, TOMORROW

TOREKELDI SHARMANOV
President of the Kazakh Academy of Nutrition, Almaty

Until the XX century high newly born infants' and children's death rate was normal. Due to the unsaturated and poor nutrition the low height, the diseases like measles, smallpox, and tuberculosis have been destroying the communities. Normal life expectancy in Antique Rome was 25 years, in England of XVII century – 35, of the middle of XIX century – 45. The sudden increase of life expectancy was occurred at the end of XIX – the beginning of XX century. In last 80 years it has grown in Chile for 2.5 times, and in the USA – 1.5 times. In China in second half of XX centuries it has increased for two times, while in the Southern Africa countries – only for 15%.

This growth, independent of economic level, was due to eradication of smallpox, sharp decrease in diseases poliomyelitis, diphteria, hoping-cough, measles, and progress in the field of preventive maintenance and prophylactic, behavioral changes and investments in education, but it was not due to incomes growth.

The speed of Americans' wealthy level growth due to increase in life expectancy was higher than the real incomes; economic growth of Great Britain during industrial revolution was for 50% linked to feeding improvement. Between 1960 and 1990 up to 15% of total economic growth is forced by death rate decrease and the increase in life expectancy for 1 year produces the national income growth for 4%. The probability of death of age less than 5 years and risk of female death rate is linked directly to economic growth. The alimentary-dependent diseases such as cardiovascular pathology, cancer, diabete and bony rarefaction represent 60% of total deaths and it will grow up to 73% by 2020. The WHO claims that nowadays a human can live 90 years and even more without diseases.

ORGANIZATION AND EXECUTION OF SANITARY-EPIDEMIOLOGIC CONTROL FOR ANIMAL PRODUCTION PRODUCTS IN KAZAKHSTAN

ANATOLY BELONOG
Ministry of Healthcare, Committee of State Sanitary-Epidemiologic Control

The article gives review of the influence of the environment to food products contamination and impact of persistent organic pollutants on the pathology of children diseases.

The high level of nitrates contamination in republic allows us to say that in most of the regions the early vegetables and melons and gourds are the most contaminated products. Heavy metals control and control for listed pollutants and polychlorinated biphenyls and their wastes in food and water require a periodical analysis and understanding to establish the flexible planning and monitoring permitting evaluation of environment contamination situation.

INFLUENCE OF THE MAIN CONTAMINATION SOURCES ON THE FOOD PRODUCTS QUALITY

MUHTAR TULTABAYEV
The Ministry of Ecology, Kazakhstan

Strengthening of the population health by maximal possible diminution of the possible negative influences by harming environmental factors is one of the most important problems of the state. Population supply with environmental pure production with high taste and nutritional value is one of the state priorities.

The last years' researches proved negative impact of the air, soils and water pollutants on the environmental characteristics of the food products. The environment quality deterioration in Kazakhstan is caused by the various factors. One of them is environment pollutions caused by oil and gas industry. For example, the results of the morphological studies of the animals and fishes identified high concentrations of heavy metals and oil products in their tissues and organs.

The next very important factor is pollutions by the former nuclear sites. Nowadays the water supplying horizons represent considerable danger for animals, soils, plants and consecutively for a human. It is identified that gamma-radiation level from the soils and dusts at the certain areas is more than 100 mR/h while the desired level is 8–10 mR/h.

Thus, population supply with environmental pure production with high taste and nutritional value is very important and necessary task.

STRATEGY OF SANITARY-EPIDEMIOLOGIC SERVICE FOR POPULATION PROTECTION FROM POLLUTANTS

K.S. OSPANOV
Republican sanitary-epidemiologic station, Almaty, Kazakhstan

Due to the problem of environment contamination, the food products may contain the different contaminants having potential danger for a human health.

The development of systems and methods of safety provide is an important state goal. The state offices of sanitary-epidemiologic control achieve the per-manent control of the objects producing and selling the animal products. The special attention is paid for the organization and execution of the laboratory methods for pollutants identification in food.

Ministry of Health of the Republic of Kazakhstan gives attention to strengthening the material and technical base. According the State Program "Reforming and development of healthcare in Kazakhstan for 2005–2010" there were made many actions on modernization of all laboratories of sanitary-epidemiologic expertise. The equipment corresponding to the international standards was bought for those purposes.

The sanitary-epidemiologic service has 217 sanitary-hygienic laboratories, centers of sanitary-epidemiologic expertise which do monitoring of raw and food products contamination by the most dangerous pollutants such as salts of heavy metals, nitrosamines, mycotoxines, pesticides, antibiotics and others.

Republican sanitary-epidemiologic station has nine profile laboratories. The newly bought equipment is used for complicated arbitrate and certification studies, of the separated cultures, decrypting the etiology of epidemic diseases and for training of the specialists in sanitary-epidemiologic expertise, getting the practice experience and skills for new equipment and methods.

The employees of the Republican sanitary-epidemiologic station and oblasts centers of sanitary-epidemiologic expertise have the educational and training programs in different countries abroad (Russia, USA, Egypt, Canada, Malaysia, China, etc.). The Republican san-epid station has created in Kazakhstan the only laboratory for asbestos studies corresponding to international standards. The new equipment and apparatus allowed us to implement new modern technologies in laboratory services.

THE ROLE OF PUBLIC ORGANISATIONS IN REALISATION OF STOCKHOLM CONVENTION ON POPS (PERSISTENT ORGANIC POLLUTANTS) AND OTHER IMPORTANT PROCESSES OF CHEMICAL SAFETY

LUDMILLIA ASTANINA

Greenwomen Ecological News Agency, Almaty, Kazakhstan

Abstract: The main part of the POPs in Kazakhstan is constituted of pesticides. The industrial POPs are obtained and used on the factories of energy production, oil treatment and chemistry production. In Kazakhstan large quantities of pesticides, inappropriate conditions of storage and packaging, the possibility of unauthorized access and non-controlled use of packaging, the big risks for human health and environment, especially at the moment of natural incidents, all those factors raise the problem of including the pesticides into the list of environmental and social problems of high priority and they demand efficient and fast decision.

Nowadays the countries of Central Asia suffer the same problems of chemical contamination and their negative influence onto the human health and environment. The Central Asia countries have a stocks of old and useless pesticides and at the same time there are no reliable monitoring and environmental control systems, also there are no burial grounds corresponding to the qualification requirements.

As the Stockholm Convention supposes the public participation in toxic substances problems the environmental NGO have accumulated certain experience in advancing the Stockholm Convention, making some informative actions and campaigns and executing some projects. Now Kazakhstan, Kyrgyzstan and Tajikistan are the members of the Stockholm Convention. This article describes the experience of Environmental Analytical Agency "Greenwomen" and other NGO of the public participation, the principles of cooperation with the government offices.

Keywords: Chemical pollution, public participation, contamination control policy

1. Introduction

Persistent organic pollutants (POPs) among other chemical substances are especially dangerous in Kazakhstan. They are heterogeneous group of chemical agents. POPs are highly toxic and even very low concentration may harm wild nature and human health. Chemical compounds and mixtures of this group are air-, water- and migrating animal-borne, as well they might precipitate at large distance from the emission point, accumulating in land and water ecosystems. Currently Central Asian countries face the same problems related to chemical pollutants and their deleterious effects on human health and environment.

Republics accumulated stocks of fusty and unusable pesticides (chemical agents used against plant pests and for extermination of weeds). There are no reliable systems on monitoring and ecological control of use and import of harmful toxic agents, there are no or insufficient number of ranges-burial grounds meeting qualification requirements for burial of harmful chemical agents.

In Kazakhstan there are about 25 million hectares of plough-land and until 1990s pesticides were used all over these lands. The total annual volumes of pesticides made 35,000–40,000 t. In 1986–1995 the volumes of chemical plants protection reduced to 1800 t. The pesticide load on 1 ha of ploughed field also reduced. Since 1998 pesticide volumes increased and currently make 9,000–11,000 t. Herbicides and fungicides compose the major part of plants protection.

In spite of the fact that in USSR DDT (dichlorodiphenyltrichloroethane) was forbidden in 1971, it was used in Kazakhstan in veterinary and medicine till 1990s.

In 1985 DDT and DDE was found in water along the piece of Syr-Daria River from boundary allotment with Uzbek SSR till Kazalinsk Town. By that time deaths of birds and fish was reported there. In the bodies of died fishes and birds was found DDT and its metabolites.

In 1982–1987, 14 cases of fish deaths were registered on the territory of Kazakhstan. Those deaths occurred due to accumulation of chlorine organic pesticides in water reservoirs. Thus, in 1987 DDT was found in one third of examined reservoirs: in water, aquatic vegetation, in invertebrate organisms, in internals of fish, in bottom sediments.

As for contamination of soils in Kazakhstan: the mean value of DDT residual quantities fluctuated from 1.2 to 5.9 MPC. In 1994, 12,000 of soil samples were taken; of them tenth part was contaminated with chlorine organic preparations. In 1993, this indicator reached one fifth. Based on this we might conclude that 10–20% of soils are contaminated with chlorine organic pesticides, with possible DDT and other SOP-pesticides presence.

Chlorine-containing pesticides are notable in the list of forbidden pesticides. These are aldrin, dildrin, DDT, heptachlorine, GHCG, polychlorinepinen, polychlorinekamphen. In the Republic the number of disabled pesticides grows every year, the number of rendered harmless preparations and package, of course, decreases.

Fusty pesticides on the territories contaminated with salt of heavy metals and radioactive nuclides are of major concern. Among stocks of unused pesticides there are preparations used in agriculture over 40 years ago.

Large quantities of unwanted and fusty pesticides along with improper storage and package conditions, possibility of unauthorized access and uncont-rolled use of package for domestic purposes, large risks for human health and environment, especially during natural disasters and man-caused incidents (floods, fires, catastrophes, etc.) raise the problem of inclusion the fusty pesticides into the list of priority ecological and social problems in Kazakhstan, requiring immediate and effective solution, in particular, soonest ratification of Stockholm Convention in Kazakhstan. As Stockholm Convention presumes involvement of public into issues related to toxic substances, since 2002 ecological NGOs of Kazakhstan perform certain activity for "Greenwomen".

"Greenwomen" participates in international processes on chemical safety, in particular, in the projects of IPEN – International POPs Elimination Network, Women in Europe for a Common Future – WECF. In 2004 International POPs Elimination Network – IPEN in cooperation with UNIDO – UN Industrial Development Organization and UNEP – UN Environment Program and under financial support of Global Ecological Fund (GEF) initiated global International POPs Elimination Project (IPEP).

Central Asian non-governmental organizations participated in International POPs Elimination Network Project in EECCA Region (Eastern Europe, Caucasia, Central Asia). Coordinator of International POPs Elimination Network Project in EECCA Region is Ola Speranskaya ("Eco-Concord" Center, Russia).

2. Experience of Central Asia: The Role of Community

Since June 2002 Project on raising awareness of State structures, NGOs, mass media and other stakeholders on POP problem was implemented in Central Asian countries. POP experts from Central Asia, NGO Environmental Analytical Agency "Greenwomen" (Kazakhstan), Center "Gender: innovations and development" (Uzbekistan) participated in the Project. Regional Ecological Center of Central Asia (REC CA, Kazakhstan) funded the Project.

Different activities were implemented in the frame of the project in Central Asian countries. In Kazakhstan Project's activity was focused on the following:

- Development, editing and dissemination of POP materials, their adaptation for information campaigns and roundtables in CA. Environmental Analytical Agency "Greenwomen" has developed special issues of "Terra – Zher Ana" Magazine devoted to persistent organic pollutants.
- "Greenwomen" has developed recommendations on conducting information campaigns for CA consultants and coordinators. In particular, the information about state structures, which might possess POP data, was presented. For effective information campaigns in CA the list of topics (issues) as well as information to be highlighted by mass media, was proposed.
- "Greenwomen" developed and disseminated the informational kits about POP problem in Russian and English for the State structures, mass media, NGO CA, international organizations, international POP networks.
- Providing consultation services for participants and target groups.
- Development of Roundtable agendas, providing methodical assistance for the RT.

The ideology of conduction the Roundtables in CA was developed; necessary visual aid was provided. The series of slides on community actions related to fulfilling commitments under Stockholm Convention and POP elimination was developed. Further this approach was successfully used in the activity of Environmental Analytical Agency "Greenwomen" and other CA non-governmental organizations.

In the frame of the project non-governmental organizations held the Roundtable "Persistent Organic Pollutants: why they cause concern, international initiatives, community actions" in Bishkek (Kyrgyzstan).

Parliamentarians, representatives of international and non-governmental organizations, ministries and agencies, mass media, POP scientists and experts participated in the Roundtable. Specialists had lectures on POP. Afterwards the small groups' discussions on identification of community's role in fulfillment of commitments under the POP Stockholm Convention were arranged. Proposals on POP problem in Kyrgyzstan were developed with further discussion in Almaty at general consultation seminar.

Informational campaigns were conducted for NGOs of Issyk-Kulskaya, Narynskaya, Chuiskaya, Talasskaya, Oshskaya and Dzhelal-Abadskaya Oblasts, mass media representatives, teachers of ecology of schools and higher educational institutions of Bishkek; information obtained from Project organizers, reports of Kyrgyzstan specialists, press-release, outcomes of RTs, Internet materials were disseminated. Information about POP was published in mass media.

Ecological organization "For the Earth" (Tajikistan) prepared kits of POP materials for NGO resource centers in Dushanbe, Kulyab, Horog and Hudjand as well as for ecological organizations and mass media.

General information about POP problem was incorporated into electronic bulletin published by "For the Earth" organization (For the Earth! Newsletter № 39 on 15 September 2002) disseminated to dozens of ecological nature-conservative organizations in CIS (including Kyrgyzstan, Kazakhstan, Russia, Turkmenistan). In the frame of informational campaign "Asia Plus" and "Sadoi Dushanbe" radios broadcasting for the whole Republic had the number of transmissions on POP problem.

The meetings with population in the Republic newspaper journalists were arranged. POP materials were disseminated among participants of the First National Conference of non-governmental ecological organizations held at the beginning of October in Dushanbe.

Information about POP problem in Turkmenistan was prepared and disseminated among state structures and specialists, non-governmental Turkmenistan organizations, all partners in CA. Information campaigns in Kazakhstan, Kyrgyzstan, Tajikistan and Uzbekistan included information about POP problem in Turkmenistan.

In Uzbekistan in the frame of information campaign and in order to attract the attention to POP problem, meetings were arranged with newspaper and TV journalists. Information packages were disseminated among state structures, mass media of the Republic, NGO and international organizations (UNDP, Doctors without limits, UNICEF and others).

POP information is as well disseminated at international conferences and workshops; by e-mail and via Internet. The exhibition of children's posters on "POP" was held. Action in the Museum of Amir Timur was arranged during International conference on reproductive health. The work of non-governmental organizations in CA was discussed in Almaty during sub-regional seminar. Recommendations to the governments were developed based on the outcomes on this work.

3. Global Day of Actions Against POP

Environmental Analytical Agency "Greenwomen" became a member of IPEN (International POPs Elimination Network), where we implemented a few small projects, for instance annual Global Days of Actions against POP. In 2005, the Global Days of Actions against POP was dated for the Day of Earth on 22 April. "Greenwomen" proposed to provide support to this action in Kazakhstan.

In the frame of this action we held press-conference in Kazakhstan press-club (Almaty), where representatives of Republican Sanitary-Epidemiological Station (RSES), Kazakhstan Business Council for Sustainable Development (KBCSD) and non governmental organizations participated.

The seminar "POP – danger for the Planet" was held in Kostanai City on the basis of Research Center on the Problems of Ecology and Biology of Kostanaiskiy State Pedagogical Institute (RC PEB KSPI), under organizational support of NGO "Naurzum". Students, graduators and academics of the Faculty of Natural Sciences participated in the seminar.

Cultural and ecological association "Boomerang" disseminated POP information in the Ridder City (Eastern Kazakhstan Oblast) for teachers of ecological disciplines. POP information and press-releases were given to different ecological organizations of Kazakhstan for conduction of information campaigns:

–   Special POP bulletin
–   Colorful leaflet "We are against of POP"
–   Press-release in Kazakh and Russian languages for mass media

Appeal from non-governmental organizations on promoting Stockholm Convention and information was given to the State ecological organizations.

4. The Future Without Toxic Substances

In 2007, Environmental Analytical Agency "Greenwomen" under support of the Program on Chemical Safety of the Center "Eco-Concord", (coordinator of International POPs Elimination Network Project in EECCA Region Olga Speranskaya, Russia) implemented Project "The Future without Toxic Substances!" In the frame of the

Project sub-regional seminar "The Future without Toxic Substances (Affect of toxic chemical substances on environment and human health in Central Asia: the ways out of problem)" was conducted. During the seminar the international experience and training on development of practical skills on public inventory of fusty and forbidden pesticides stocks and methodical guidelines were presented. Possibilities of CA countries to realize the objective **"The year 2020 without toxic substances!" were analyzed**. The current situation with participation of community in the decision-making, review of situation on introduction of СПМРХ principles and appropriate conventions (Stockholm, Rotterdam, Basel etc) into legal acts of CA was evaluated. IPEN representatives, IPEP partners, International Chemical Secretariat, WECF presented their experience and projects on solving problems related to affect of toxic chemical substances on human health and environment. Currently, the Project is on-going; one of its results will be regional approach (or plan of action) to implementation of objective **The Year 2020 – without toxic substances.**

5. Recommendations of Non-Governmental Organizations to the Governments of CA Countries

"Greenwomen" as other non-governmental organizations is against toxic substances and calls the governments of our countries to active and consistent implementation of global ecological objective – achievement of rational regulation of chemical substances to 2020. This objective was set in 2002 by delegates of World Summit on Sustainable Development (Johannesburg, SAR).

The community insists the governments of CA take measures to ensure the chemical substances produced and used in a way to prevent significant affect on human health and environment.

Objective – achievement to 2020 of rational regulation of chemical substances – is essential for our countries.

Proposed **"Global Plan of Actions"** in the frame of СПМРХВ strategy emphasizes the importance of participation of all concerned parties in it. Plan also indicates an importance of maximal transparency and openness during strategy implementation as well as participation of community in taking key decisions.

The governments shall:

- Ensure effective participation of community in regulation of chemical substances (policy, legislation, plans of actions, programs, projects develop-ment)
- Ensure full and timely community information, hold consultancies with all concerned parties, ensure public opinion considered in decision-making process
- Arrange educational programs for representatives of state institutions, non-governmental organizations, local governments, heads of enterprises, journalists on chemical safety
- Pay special attention to the dialogue between representatives of industrial plants, state environmental institutions and community
- Develop collaboration with international organizations in the area of chemical substances regulation

Kazakhstan as well as other CA countries accumulated insufficient experience of participation in ecological decision-making process. We should continue developing basic issues of community participation in ecological sphere.

L. ASTANINA

References

Information provided by Aiman NAZHMETDINOVA, head of the Department of Pesticide Toxicology of the Republican Sanitary-Epidemiological Station (RSES), Ph.D. situation analysis on POP in RK in the frames of UNDP/GEF Project "Initial assistance to RK in fulfilling commitments under the Stockholm Convention".

Materials and Information of sub-regional seminar "The Future without Toxic Substances! (Affect of Toxic substances on environment and human health in Central Asia: way out of problem)", Almaty, 2007.

PART I

Characterization of the environmental failures in relationship with human and animal health

STUDY OF COMPOSITION AND PROPERTIES OF OIL POLLUTION

YERDOS ONGARBAYEV, ZULKHAIR MANSUROV
Al-Farabi Kazakh National University, Almaty, Kazakhstan

Abstract: The composition and physico-chemical characteristics of oil pollution are studied to develop an effective technology of recycling. The content of an organic phase varies from 5 up to 95 mass% and its content depends on the initial spilled oil. However, the content of high-molecular compounds in composition of an organic phase increases due to the effect of atmospheric factors. Study of oil pollution composition and properties will enable to reduce the existing accumulations of oil wastes and their harmful influence on the environment.

Keywords: Oil pollution, oil wastage, oil contaminants

1. Introduction

The discovery of large natural reserves of hydrocarbon raw materials and development of the oil-and-gas industry, use of the richest stores of power and building materials have changed in a short period of time the shape of waterless desert and have simultaneously created ecologically crisis sites for vital functions of biogeocenosis and social and economic tension in the oil-production regions.

Now normalization of ecological conditions and rational use of natural resources have become the major state problem of the region. In this connection, the duly estimation of an ecological condition of territories, physical and chemical researches of composition of oil wastes, study of processes of anthropogenic changes of soils, the content of heavy metals and hydrocarbons in them, as well as development of practical recommendations on the decrease in anthropogenic pollution of the territory, methods of recycling of industrial wastes at the enterprises of the oil-and-gas complex are actual (Wang et al., 1999).

Petrochemical pollution of soil is marked on the area of all operating oil-and-gas deposits and is connected with irrational use of natural raw material resources, the out-of-date and worn out production equipment, high paraffin composition of hydrocarbon raw material (Cole, 1994). The basic sources of pollution thus are crude oil and industrial sewage, sulfur and nitrogen oxides, hydrogen sulfide, phenol, ammonia, gas and oil sludge (Barcelona and Robbins, 2002).

Oil consists of a great amount of hydrocarbons and high-molecular tar and asphalt-tene substances. Its main components are carbon (83–87%), hydrogen (12–14%) and oxygen (1–2%), as well as nitrogen, sulfur, various microelements (Xie and Barcelona, 2001). Saturating a soil profile, oil renders a negative influence on the growth and development of plant by direct influence of naftenic acids and other hydrocarbons and mineral salts. Besides, crude oil includes carcinogenic polycyclic hydrocarbons (naftalans, acenaftens, fluorens, pyrene and benzpyrene), causing malignant tumors causing in a living organism (Heath et al., 1993). In the process of migration in a chain soil – plants – living organisms, aromatic hydrocarbons cause radical changes in brain cells, composition of blood and blood forming organs, vegetative infringement.

In pollution, the presence of a great amount of paraffin, tars and asphaltenes in oil renders a considerable negative influence on physical and chemical properties of soil and the condition of the environment (Einhorn et al., 1992). Paraffin (C_{10}–C_{34}) is a primary product of lipid fractions of an initial biomass, it is not toxic for plants, but at low temperatures it crystallizes into a strong mass, adsorbs tar substances and asphaltenes, thereby essentially influencing moisture-air exchange and soil compaction forms bitumen crusts in the profile impregnated by black oil and chemical reagents. A short-term spouting or jet spill of oil at wells and oil pipelines cause the formation, in the soil profile, of fine surface-bituminous crusts which usually quickly degrade and collapse. In oil "barns" and at long solidification of great volumes of crude oil on the surface powerful bituminous crusts are formed (thickness of 20–30 cm and more). They differ in high density (volumetric weight of 1.8–1.9 g/cm^3, the parameter of coupling up to 80–90%), are poorly oxidized on air, are poorly accessible to microorganisms, therefore slowly decay and are long kept in the soil profile.

Thus, soil becomes the basic source of uptake of radioactive elements into plants and further, via a food chain, into animals and man causing genetic, somatic and oncological diseases. This determines the necessity of duly deactivation of soils and the production equipment, realization of monitoring observation of ecological conditions (Meegoda and Ratnaweera, 1995).

The technology of oil production entails great difficulties in paraffinization of the underground and overground production equipment, scaling in face zones and communications, drowning oil wells and corrosion of the equipment. All this results in frequent breaks of pipes, discharge of crude oil and mineral sewage onto the surface of soil, formation of peculiar geochemical fields with the high contents of hydrocarbons (Mansurov et al., 2000, 2001, 2004). Emergency discharge of oil at rocker frames is made into barns. Accumulation of oil contaminated soil is stored on special ranges representing diked foundation ditches of a great capacity. On the Ozen deposit (Kazakhstan) 43 barns are placed on a total area of 2,183 ha (Ongarbayev and Mansurov, 2003). The largest of them reaches 70 ha at average depth of 2.5 m (Table 1).

TABLE 1. The characteristics of oil barns

Deposit	The area (m^2)	Thickness of an oil layer (m)	Volume of oil (m^3)	Quantity of oil (t)
Ozen	600	0.6	360	289
–/–/–	900	0.9	720	536
–/–/–	1,200	0.6	720	367
–/–/–	1,800	0.8	1,440	1,156
Zhetybai	840	0.4	336	270
–/–/–	680	0.2	130	105
–/–/–	800	0.3	240	195
–/–/–	7,000	0.1	700	568
–/–/–	12,000	0.1	1,200	975

For scientific substantiation and determination of chemical influence of components of crude oil, oil wastes and technological solutions on the environment under various conditions of their storage and meteorological conditions, as well as at external influence of sunlight and new portions of crude oil on oil wastes of the salted ground complex physical and chemical methods of research were carried out.

2. Materials and Methods

The objects of researches were: barn oil of Ozen deposit (Mangistau area, Kazakhstan), oil sludge after clearing tanks and repair of the equipment and the ground polluted with oil, spilled as a result of the break of Ozen-Zhetybai-Atyrau main oil pipeline. For extraction of an organic part from oil waste the extraction method was used. The ethanol and benzene mixture (the ratio of ethanol: benzene was equal to 1:4) was used as a solvent. Extraction was carried out in Soxhlet apparatus till termination of solvent coloring.

The distribution of hydrocarbons of normal and isoprene structures in the organic part of barn oil was investigated by the method of gas chromatography. The gas-liquid chromatographic analysis was carried out on chromatograph "Chrom-5" at a step programming of temperature from 230°C up to 560°C with the use of a capillary column filled with apezone L, and hydrogen as a gas-carrier. The group composition of the organic part of oil waste was determined by the changed technique of VNIINP based on the adsorption-chromatographic method of Marcusson, based on sedimentation of asphaltenes by petroleum ether. For separation of oils and tars, their different abilities to be sorted by a silica gel were used.

The fractional composition of the organic part of oil waste was determined by distillation under vacuum at pressure of 4 mmHg in ARN-2 apparatus.

The content of microelements in the organic part of oil waste was determined by the method of emission spectral analysis with excitation of spectra in an arc of a direct current. The spectra were registered on diffraction spectrograph DFS-13 at the strength of current 10 A and a diaphragm 5.

Infra-red spectra of the organic part of oil waste were taken on two-beam automatic spectrometer UR-20 at 400–4,000 cm^{-1}. A folding dish with plates from potassium bromide was used; thickness of an absorbing layer was 0.01 mm. ESR spectra of the organic part of oil waste were studied at room temperature in an atmosphere of air on radio spectrometer RE-1306. The spectra were taken on frequency of 9,400 MHz at intensity of the magnetic field 3,350 ersted. To determine the concentration of paramagnetic centers, Mn^{2+} deposited on MgO was used as a reference.

Roentgenograms of initial oil wastes and their mineral part were made on diffractometer DRON-3M at an accelerating pressure 35 kV with the use of tubes with a copper cathode (a nickel board). The samples were prepared by pressing the portions of the same weight into a special dish. Registration of roentgenograms in the range of angles from 2° up to 40° was carried out with the rate of 2°/min in a mode of continuous scanning recording the curve on a tape.

3. Results and Discussion

The phase composition of the investigated oil wastes by the results of extraction is presented in Table 2. The content of an organic part in oil sludge and barn oil makes up 81.8 and 84.9 mass%, and in the oil contaminated soil – 27.2 mass%. The content of water in oil sludge and barn oil is 8.1 and 7.1 mass% accordingly, while in the polluted ground it is only 1.0 mass%. The basic part of the polluted ground consists of the mineral part (71.8 mass%), its content in oil sludge and barn oil is 10.1 and 8.0 mass%.

TABLE 2. Phase composition of oil wastes

Component	Oil sludge	Barn oil	Oil contaminated soil
		The content (mass%)	
Organic part	81.8	84.9	27.2
Water	8.1	7.1	1.0
Mineral part	10.1	8.0	71.8

3.1. BARN OIL

At the enterprises of a joint-stock company "Ozenmunaigas" as a result of a purge of the equipment on group installation (GI), when working on wells and on waste collectors the wastes containing barn (pit) oil are formed.

Complex research of the given type of wastes has shown that its molecular composition depends on the place of sampling. For example, the average part of barn oil does not differ from crude oil in composition. Oil which is closer to boards, storehouses (barn) is polluted by soil components. Therefore, a great attention is paid to the research of the ground composition and evaporation process of hydrocarbons from wastes.

Physical and chemical characteristics of oil from barn GI 57 Block 3A of Ozen deposit have been investigated. Unlike crude oil, barn oil does not contain light fractions, is characterized by high density (950–1,000 kg/m^3 and more) and extreme heterogeneity of hydrocarbonic and componential compositions. Samples of an organic part of Ozen deposit barn oil have heavy dense weights and are characterized by a high enough content of paraffins of a high molecular weight, and also high values of viscosity (>500 mPa·s).

A great amount of heavy oil residues in waste contributes, first, to a sharp decrease in mobility of oil under the influence of gravitational surface and capillary forces, secondly, form a dense impenetrable film on a sole and walls of a barn. Therefore migration of toxic components from a barn into underground waters can be neglected.

The crude oil stored in ground barns is exposed to various external influences (ultraviolet radiation, change of temperature, pressure, etc.), therefore, its chemical composition and properties change. Thus favorable conditions for formation of various classes of volatile, toxic and chemically active hydrocarbons are created. The structure of crude oil is stored in barns is investigated by means of the method of gas chromatography.

The method of gas-liquid chromatography allows classifying the organic part of oil waste, as well as oil by chemical types. By the relative content of normal and isoprene alkanes the organic part of oil waste can be referred to one of oil types: A^1 and A^2 (paraffinic), B^1 and B^2 (not containing n-alkanes). Type A^1 includes light paraffinic oil in which the content of n-alkanes C_{17} and C_{18} prevails over the content of pristane (C_{19}) and fitane (C_{20}) (oil biomarkers).

Figure 1 presents the chromatogram of barn oil. It is seen that barn oil refers to type A^1. Despite the absence of light fractions (up to 200°C), that is characteristic for all oil wastes, in the given sample there are n-alkanes C_{15}–C_{34} and isoprenoids C_{14}–C_{20} with obvious prevalence of n-alkanes C_{17} and C_{18} over pristane (P) and fitane (F). The chromatogram is distinguished by the presence of solid n-alkanes above C_{30} in sufficiently high concentrations that indicates the significant content of oils and solid paraffins in barn oil.

Besides, in the chromatogram there is a continuous naftene background, indicating the reference of the given sample to oil type B[1]. The given background is characteristic for extracts of oil bitumen rocks lying close to a day time surface, in the conditions favorable for chemical and biological oxidation of hydrocarbons.

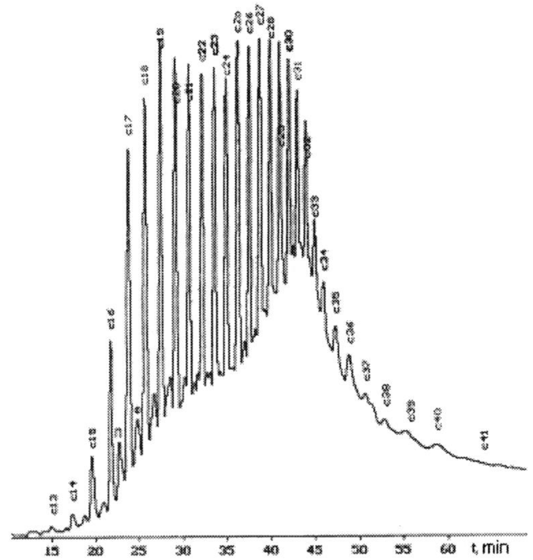

Figure 1. Chromatogram of an organic part of barn oil

According to the chromatographic analysis (Table 3) the organic part of barn oil has appeared to be high paraffinic. In the range of temperatures 270–470°C concentration of paraffins makes up 1.6–2.2%.

The greatest phase transitions are observed at the change of temperature from 288°C up to 415°C. The marked effect indicates a quasi-phase transition or the beginning of a phase transition in the oil system under consideration in the range of 280–400°C, i.e. with formation of phases at sufficiently low temperatures that allows to assume, that molecules of paraffins do not enter the composition of the complex structural units responsible for the process of formation of carboids and carbenes.

TABLE 3. Hydrocarbonic composition of an organic part of Ozen deposit barn oil according to the chromatographic data

C_{13}–C_{22}										
C_n	C_{13}	C_{14}	C_{15}	C_{16}	C_{17}	C_{18}	C_{19}	C_{20}	C_{21}	C_{22}
T (°C)	235	254	270	287	303	317	331	345	357	368
C (%)	0.036	0.455	1.69	3.54	8.45	8.55	7.95	7.74	6.72	5.77

C_{23}–C_{32}										
C_n	C_{23}	C_{24}	C_{25}	C_{26}	C_{27}	C_{28}	C_{29}	C_{30}	C_{31}	C_{32}
T (°C)	380	391	403	414	426	438	448	450	460	470
C (%)	6.16	4.98	5.76	5.37	4.78	4.36	3.90	3.26	2.34	2.18

C_{33}–C_{41}									
C_n	C_{33}	C_{34}	C_{35}	C_{36}	C_{37}	C_{38}	C_{39}	C_{40}	C_{41}
T (°C)	480	490	500	511	520	530	540	550	560
C (%)	1.03	0.934	0.921	0.949	0.438	0.301	0.570	0.590	0.201

The data of fractional composition show that there are no light fractions in it, the mass content of fractions (%): 240–340°C – 10.3; 340–500°C – 31.6; vat residue – 55.5; water – 2.6.

Thus, the analysis of the obtained data shows that Ozen deposit barn oil is a concentrate of asphalt-tar-paraffin deposits, this being indicated by the high content of solid paraffins in it.

3.2. OIL SLUDGES

Among wastes formed at the enterprises the most ecologically dangerous are oil sludges which represent a quite steady three-componental system: solid-oil-water stabilized by the presence of a gaseous phase – products of biological destruction of organic substances.

As a result of natural evaporation of moisture in sludge stores the liquid part of oil sludge is lost. Viscous, sticky paste of a different degree of plasticity is formed. According to laboratory researches at humidity 30 mass% oil sludge represents a fragile tile which can be transported by a motor transport. In such condition oil sludges do not dust, possess strongly pronounced thixotropic properties; give off no more than 1–3 mass% of moisture to the underlying layers of ground. Thus, dry oil sludge is convenient and safe for transportation and keeping in storage. By toxicity, oil sludges are industrial wastes of the 3rd class of danger and are one of the sources of toxic heavy metals, and are distinguished by high radiation.

Physical and chemical characteristics and element composition of an organic part of oil sludge and oil contaminated soil are represented in Table 4, whence follows, that an organic phase of oil waste – heavy (989 and 996 kg/m^3), sulphurous (the content of sulfur 1.0 and 2.1 mass%), is characterized by high parameters of coking capacity and acidity. The acid number allows judging the quantity of oxygenous components in an organic part of oil sludge. Its high value in an organic phase of oil sludge, despite the primary content of paraffin-naftene hydrocarbons in it, can indicate the presence of significant intermolecular interaction between oil sludge components that imparts to its high adhesive properties.

The group composition of an organic phase of oil sludge and oil contaminated soil is presented in Table 5.

TABLE 4. Physical and chemical properties and element composition of a hydrocarbon part of the oil wastes

Parameter	Organic part of oil wastes	
	Oil sludge	Oil contaminated soil
Density at 20°C (kg/m^3)	989.0	996.0
Coking capacity (mass%)	12.5	32.1
Acidity (mg KOH/100 ml)	21.0	–
Element composition (mass%)		
Carbon	82.3	84.2
Hydrogen	11.3	9.5
Nitrogen	0.4	0.3
Sulfur	1.0	2.1

TABLE 5. Group composition of an organic part of the oil wastes

Component	Oil sludge	Oil contaminated soil
	Content (mass%)	
Oils		
Paraffin-naftene	40.8	6.5
Aromatic	26.1	28.1
Sum of oils	66.9	34.6
Tars		
Petroleum ether-benzene	11.5	35.8
Benzene	4.3	3.1
Ethanol-benzene	4.5	14.6
Sum of tars	20.3	53.5
Asphaltenes	12.8	11.9

The organic part extracted from oil sludge contains paraffin-naftene (40.8 mass%) and aromatic hydrocarbons (26.1 mass%). It is characterized by a significant content of petroleum ether-benzene tars (11.5 mass%) in comparison with benzene and ethanol-benzene tars (4.3 and 4.5 mass%, respectively). The quantity of asphaltenes in oil sludge is 12.8 mass%. Thus, the hydrocarbon phase extracted from oil sludge is characterized by the high content of paraffin and asphalt-resinous substances that strongly complicates processing of such raw material.

The data of fractional composition show, that there are no light fractions in the organic part extracted from oil sludge, the mass content of the fractions which are boiling away up to 330°C, makes up only 9.4 mass%: 240–270°C – 3.5 mass%; 270–300°C – 2.3 mass%; 300–330°C – 3.6 mass%. The output and properties of oil fractions are given in Table 6. As is seen from the structural-group analysis, oil fractions have the paraffin-naftene nature and average value of density.

TABLE 6. The output and properties of oil fractions of the oil sludge organic part

Parameter	Oils by fractions (°C)		
	240–270	270–300	300–330
Output of oils (mass%)	3.5	2.3	3.6
Density (kg/m^3)	831	845	853
Parameter of refraction n_D	1.460	1.467	1.471
Molecular weight	193	218	247
Structural-group composition: the content of carbon (%)			
In aromatic rings C_{Ar}	6.6	6.9	6.2
In ring structures C_R	47.6	48.5	46.1
In naftene structures C_N	41.0	41.6	39.9
In alkyl substituents C_A	52.4	51.5	53.9

The results of the spectral analysis (Table 7) have shown that in an oil sludge organic phase the content of copper, chrome and manganese is high, and high concentration of vanadium, iron, magnesium, silicon and aluminium is characteristic for an organic part of the oil contaminated soil. The content of nickel in an organic phase of both wastes is equal. The level of the content of all metals exceeds maximum permissible concentration (maximum concentration limit) in soil.

TABLE 7. The content of microelements in the organic part of oil wastes

Metal	The content in organic part (mass%)		Maximum concentration limit in soil (mass%)
	Oil sludge	Oil contaminated soil	
Vanadium	≤0.003	>0.01	$5 \cdot 10^{-5}$
Iron	≥0.001	~0.1	–
Nickel	<0.003	0.003	$4 \cdot 10^{-6}$
Copper	<0.3	0.003	$3 \cdot 10^{-6}$
Chrome	~0.03	<0.001	$6 \cdot 10^{-6}$
Manganese	≤0.01	0.003	$7 \cdot 10^{-4}$
Titanium	–	0.003	–
Magnesium	–	≥0.01	–
Aluminium	+	~0.01	–
Silicon	+	0.03	–

The data of the IR-spectroscopic analysis have shown that in chemical structure of an oil sludge organic part there prevail the saturated structures in the form of groups CH, CH_2, CH_3 (absorption bands at 3,000–2,800, 1,465, 1,380 cm^{-1}) and groups CH_3 which are a part of long paraffin chains (720 cm^{-1}). Besides, the presence of aromatic structures (1,600 cm^{-1}) and a weak absorption band of oxygen containing functional groups (1,700 cm^{-1}) are observed.

Such factors as thermal and solar irradiation, chemical and mechanical influence cause the break of chemical bonds and formation of free radicals in molecules of an organic part of wastes. In oil wastes there are long-living and short-living free radicals. The existence of long-living radicals is caused by the presence of the complexes containing an ion of vanadium (4+), as well as free bonds in the condensed aromatic structure of tar-asphaltene parts the quantity of which is proportional to the content of asphaltenes. Short-living free radicals appear in an organic phase and in chemical reactions, their concentration can serve as a direct characteristic of the organic phase reactivity.

A singlet of free radicals of a sufficiently high intensity and the fourth component of the super fine structure of vanadium (4+) are registered in the a spectrum of oil sludge. The concentration of free radicals makes up the order of $8.6 \cdot 10^{17}$ spin/g, and that of vanadium – $1.4 \cdot 10^{17}$ spin/g. Vanadium is a part of an organic part in the form of a vanadilporphirine complex.

According to the X-ray phase analysis data, the oil sludge organic part is represented by polynaftene ($d_{ool} = 4.66$ A) and graphite phases ($d_{ool} = 3.355$ A). In the roentgenogram of oil sludge there are also very intensive peaks corresponding to hexahydrate of aluminium mellitate $C_6(COO)_6Al_2 \cdot 6H_2O$ ($d_{ool} = 4.129$; 3.704; 2.478; 2.209; 2.087 A) the presence of which can be accounted for the addition of active reagents into oil for improvement of its transportation by pipelines. According to the carried out researches, the basic components of the mineral part are quartz (52%), calcite (15%), halite (16%), albite (7%), adularia (5%) and hematite (5%).

According to the gravimetric analysis the mineral part of oil sludge consists of sand with the following grain structure (the average values), mass%: fractions larger than 0.63 mm make up 4.13; fractions of 0.4–0.63 mm – 39.35; fractions of 0.25–0.4 mm – 15.32; fractions of 0.1–0.25 mm – 11.53; fractions less than 0.1 mm – 29.67. By the grain composition the oil sludge mineral part refers to fine-grained.

Thus, the analysis of the obtained data shows that by the composition and properties, the oil sludge organic part comes closer to the heavy residues. It is characterized by high contents of tars and asphaltenes, high density due to the effect of climatic factors on the wastes during a long period of storage in open storehouses.

3.3. THE OIL CONTAMINATED SOIL

The oil contaminated soil, basically, is formed as a result of liquidation of old spills of oil and has the content of an organic part up to 30 mass%. By virtue of physical properties of oil its penetration into soil does not exceed 5–7 cm. In this connection an elastic water-proof substrate is formed which does not render any influence on soil and ground.

The analysis shows that samples of the oil contaminated soil are non-uniform in the content of hydrocarbons (Table 5).

The organic part of the ground polluted with oil is represented by insignificant quantity of paraffin-naftene oils (6.5 mass%), but the high content of aromatic oils (28.1 mass%). In comparison with oil sludge it contains a lot of tars: petroleum ether-benzene 35.8 mass%, ethanol-benzene – 14.6 mass%, benzene – 3.1 mass%. The content of asphaltene is high – 11.9 mass%. By these parameters the hydrocarbon phase of the ground polluted with oil can be considered high viscous.

The roentgenogram of the oil contaminated soil shows that its mineral part includes quartz SiO_2 (d_{ool} = 3.338 A), calcite $CaCO_3$ (d_{ool} = 3.025 A) and feldspar and in the subordinated value – micas, clay minerals (chlorite and kaolin $Al_4Si_4O_{10}(OH)_8$).

When studying paramagnetism of the oil contaminated soil at room temperature, a fine structure of six lines is observed in ESR spectra . The width of lines corresponds to the width of signals of the standard of g-factor Mn^{2+}/MgO, hence, this structure refers to Mn^{2+} ions. Probably, Mn^{2+} ions are in the matrix of $CaCO_3$ (shell rock) of a waste mineral part. Their concentration is estimated by the order of $1.2 \cdot 10^{17}$ spin/g. Besides manganese, the soil contains a significant amount of ferriferous impurities which is indicated by a wide line of a ferromagnetic resonance.

In the central part of ESR spectrum, the most intensive line in the spectrum was observed which was referred to free "carbon" radicals the concentration of which indicates the degree of structurization of the system. This line of ESR spectrum of the sample of the oil contaminated soil represents single line with the width 5 ersted with the g-factor 2.003. Concentration of free radicals is equal to $4.4 \cdot 10^{17}$ spin/g. It is less than the concentration of free radicals in oil sludge as in the oil contaminated soil the content of an organic part is less than in oil sludge.

In the spectra one can see the growth of the peak of free radicals with the increase in concentration of an organic component in an oil waste and gradual development of a weaker peak of a vanadil-ion. The granulometric composition of a mineral part of the oil contaminated soil is as follows (mass%): 7.18 – fractions are larger than 5 mm; 13.74 – fractions of 2.5–5 mm; 9.05 – fractions of 1.25–2.5 mm; 18.0 – fractions of 0.63–1.25 mm; 19.6 – fractions of 0.315–0.63 mm; 27.9 – fractions of 0.1–0.315 mm; 4.53 – fractions less than 0.1 mm.

The use of the oil contaminated soil for obtaining of target products is complicated because of the high content of mechanical impurities basically, clay particles and the low content of an organic part.

4. Conclusion

According to the results of examination it is stated, that the content of hydrocarbons in the investigated oil wastes shows the probability of their anthropogenic action on soil, underground waters and an air atmosphere. The facts of fixation of aliphatic and aromatic hydrocarbons also confirm their potential danger for the environment. Heavy aromatic hydrocarbons can be accumulated in soils and under the first "favorable" conditions get into underground waters. Alongside with barn oil and the oil contaminated soil, another source of heavy metals and hydrocarbons is oil sludge. Therefore, it is necessary to store them in special storehouses.

The composition and structure of an organic phase of wastes change at oxidation under the influence of an atmosphere at their storage in open barns and in soil. Eventually, ageing of wastes takes place because of volatilization of light fractions. As a result of slowly proceeding oxidizing reactions of condensation and aeration of light hydrocarbons, high-molecular compounds are formed in the waste according to the scheme:

$$\text{Aromatic hydrocarbons} \rightarrow \text{tars} \rightarrow \text{asphaltenes}$$

The tendency of an organic phase to chemical transformations depends on its composition and, first of all, on the presence of easily oxidizing groups and bonds in molecules. It is well-known that in the mixtures aromatic and paraffin-naftene hydrocarbons aromatic components are much more quickly oxidized. Ageing of oil waste is strongly affected by water which contributes to acceleration of chemical transformations, to dissolution and washing out of low-molecular compounds.

References

Barcelona, M.J., Robbins, G.A., 2002, Soil and Groundwater Pollution, *Encyclopedia of Physical Science and Technology*, 15: 49–62.

Cole, G.M., 1994, Assesment and Remediation of Petroleum Contaminated Soils. CRC, Boca Raton, FL, 360 p.

Heath, J.S., Koblis, K., Sager, Sh.L., 1993, Review of Chemical, Physical and Toxicologic Properties of Components of Total Petroleum Hydrocarbons, *Journal of Soil Contamination*, 2(1): 1–25.

Einhorn, I.N., Sears, S.F., Hickey, J.C., Viellenave, J.H., Moore, G.S., 1992, Characterization of Petroleum Contaminants in Groundwater and Soils. In Hydrocarbon Contaminated Soils, Vol. 2, P.T. Kostecki and E.J. Calabrese (Eds). Amherst, Massachusetts, USA, pp. 89–143.

Mansurov, Z.A., Ongarbaev, E.K., Tuleutaev, B.K., 2000. Utilization of Oil Wastes for Production of Road-building Materials, *Eurasian Chemico-Technological Journal*, 2(2): 161–166.

Mansurov, Z.A., Ongarbaev, E.K., Tuleutaev, B.K., 2001, Contamination of Soil by Crude Oil and Drilling Muds. Use of Wastes in Production of Road Construction Materials, *Chemistry and Technology of Fuels and Oils*, 37(6): 441–443.

Mansurov, Z.A., Ongarbayev, E.K., Tuleutaev, B.K., 2004, Development of Methods of Thermal Processing and Oxidation of Oil Wastes, *Oil Refining and Petrochemistry*, 8: 49–54 (in Russian).

Meegoda, N.J., Ratnaweera, P., 1995, Treatment of Oil-Contaminated Soils for Identification and Classification, *Geotechnical Testing Journal*, 18: 41–49.

Ongarbayev, E.K., Mansurov, Z.A., 2003, Oil Wastes and Methods of Their Utilization. Kazakh University, Almaty, 160 p.

Wang, Z., Fingas, M., Page, D.S., 1999, Oil spill identification, *Journal of Chromatography A*, 843: 369–411.

Xie, G., Barcelona, M.J., 2001, Efficient Quantification of TPH and Applications at Two Contaminated Sites, *Ground Water Monitoring and Remediation*, 64–70.

THE PROBLEM OF FOOD-PRODUCTS SAFETY IN KAZAKH REPUBLIC

LUCIA I. KALAMKAROVA, OLGA V. BAGRYANTSEVA
Kazakh Academy of Nutrition, Almaty, Kazakhstan

Abstract: In spite of taking by KR government action on protection of population from low-quality and dangerous food-products, situation is not quite clear. The weight of illnesses caused by food-products which have been contaminated by various xenobiotic originated to anthropogenic and natural are state at the higher level.

In Kazakhstan the next problems of control of biology activity food supplements to nutrition are: absence of modern equipment, absence of the assortment of chemicals and standards which could help to broaden the specter of detected indexes, and absence of possibility to give full information about prepared production (nutrition and biological value, enzymes, active principle and so on). Therefore there are need to improve sanitary-hygienic demands to production of biological activity food supplements and its compositions in harmony with published rules by European Parliament (2002). For safety of population in KR, there are needs to establish the control and monitoring of genetically modified products at the markets of republic.

The major task is to improve the system of quality and safety control of food products in KR and to joint operations of different departments which must be responsible for food-staff products quality during all cycle of its preparing and consumption.

Keywords: Food-staff, pollution, quality, safety, control

1. Introduction

In spite of taking by KR government action on protection of population from low-quality and dangerous food-products, situation is not quite clear. The weight of illnesses caused by food-products which have been contaminated by various xenobiotic originated to anthropogenic and natural are state at the higher level.

It is known that the solution of problem of food-stuff quality and safety is the best way to prevent the developing of alimentary dependent pathology. Solution of this problem based on the using of different modern methods investigation of food-stuff (microbiological, toxicological, histological, sanitary-hygienically, etc.). Moreover to achieve success in this field it is necessary to join works of different government's departments in Kazakhstan and, of course, all of them must function in harmony with international standards of Codex Alimentarius, ISO, HACCP and resolution of EC. The solution of this problem is depending of the introduction of Kazakh Republic in WTO.

2. The Current Situation of Food Contamination in Kazakhstan

The priority contaminants of food-stuff in KR, like all over the world, are microorganisms and its metabolites, mycotoxins, heavy metals, nitrates, nitrites, N-nitrosamines, radioactive nuclides, some hormones, pesticides residues, polychlorinated biphenyls, dioxins, antibiotics. According with WHO data 90–95% of food poisoning cases has a microbiological etiology. At the same time in the last years the quantity of illnesses where immediate causes are opportunistic microorganisms (*Staphylococcus* sp., different *Enterobacteria, Pseudomonas aeruginosa*, Candida, other Fungi) were increased.

The **microorganisms** – different species of *Shigella, Listeria monocitogenes*, enterohemolytic species of *E. coli, Vibrio parahemoliticus, Yersinia enterolytica, Campylobacter jejunum, Aeromonas hydrophila* and *Pleziomonas shiggeloides* – could

B. Faye and Y. Sinyavskiy (eds.), *Impact of Pollution on Animal Products.*
© Springer Science + Business Media B.V. 2008

be a cause of food born infection. However these microorganisms were not included on the list of microorganisms for detection without fail in our republic.

One of the most dangerous pollutants of food-stuff is fungi and their metabolites – **mycotoxins**. They could be the cause of mycotoxicosis. The different biological effects of mycotoxines have been described – oncogenic, mutagenic, embryotoxicity, teratogenic, gonadotrophic and others. Investigations conducted by research officers of Quality and Safety of Food-Stuff Laboratory of Kazakh Academy of Nutrition during years 1985–2005 revealed fungus (*Aspergillus flavus, Aspergillus fumigatus, Pennicillium expansum, Penicillium variable, Fusarium sporotrihiella*, etc.) and mycotoxines (in 14.2% samples) in the main food-products including food for infants (milk products, dairy product, some vegetable and fruit juice, mash potatoes, pastes).

In the republic takes place the problem of monitoring of **antibiotic residues** in the food-stuffs. Especially dangerous are the antibiotics contaminating livestock production (milk, eggs, meat) because they could be the reason of allergic reactions, misbalance of gut microbiocenosis, appearing resistance to antibiotic pathogenic and opportunistic microorganisms, reduction of therapies effects of antibiotics. Our analysis of food-stuff concerning the present of antibiotic showed that penicillin residues determined in 10 milk samples (1.19%) was 0.12–1.24 standard units/ml (MPC < 0.01 SU/ml). From 195 samples of poultry and their product, 13 samples (6.6%) included antibiotics (more frequently streptomycin) from 1.24 to 9.92 SU/g (MPC < 0.5 SU/ml). In food-products for infants, meat and their products, residues of antibiotics were not detected (Hadzhibaeva, 2006).

Especially strong measures must be used in case of **pesticides residues**. More than 100 pesticides are known, but it is impossible to find all of them in food-stuff. It must be necessary to know which pesticides were used more often in other countries. For example, we detected pesticides Ramdor in white cabbage which was not included in the list of pesticides analysis without fail. As a result of our investigations some countries changed rules which regulate safety requirements to tea. The number of pesticides for analysis without fail was increased from 3 to 15 (Nazhmetdinova, 2005; Kalamkarova et al., 2004).

In Kazakh Republic, practically the problem of **genetically modified products** (GMO) was not clearly approached, in spite of definite points in the law "About safety of food-stuff" (2007). Solution of this question is important, because Kazakhstan is going to be included in WTO. Now in Kazakh Republic only one laboratory could detect GMO. It is the Department of Quality control and Food Safety of Kazakh Academy of Nutrition. The results of investigation in this field are reported in Table 1.

For successful solution of the problem of quality and safety of GMO in the Republic there is a need for developing by-laws resolutions and arrangements which could based on the "Conception about ecological safety of KR at the period from 2004 to 2015 years", and law "About safety of food-stuff" (2007), "Kartahenskiy protocol on biosafety to Convention about biological variety" (2000).

The key problem in this case is the investigation of chemical compounds of new production, which must include detection of staff quality and sanitary-hygienically data of safety. On the basis of these results, the question about registration of GMO and there broad using could be considered.

TABLE 1. The results of investigation of food-products on the presence of transgenic DNA

Name of food-stuff	Number of samples	Number of positive results
Soybean, soy isolate, soy products	94	19 (+2 traced quantity)
Sausage and sausage production	14	4
Crab's sticks	1	–
Canned meat	25	–
Bakery production	2	–
Biologically activity supplements	28	3 (+8 Traced quantity)
Nutrition for infants	4	–
Canned maize	2	–
Sum total	170	26 (+10 Traced quantity)

Organization of PCR laboratories also will give possibility to solve the problem of falsified food-staffs, because with the help of this method it could be possible to reveal solitary molecules DNA in samples.

Like other countries, Kazakhstan is confronted to the problems of control of Biologically Activity Supplements (BAS), because the laboratories don't have modern equipment, higher quality reagents and standards, and possibility to characterize nutritional and biological value, enzymes, active principle and other components of BAS. Therefore, there are necessity for improving sanitary-hygienically safety requirements of the BAS production and their compositions, in harmony with EC regulations (Directive 2002/46/EC).

At present time BAS in KR examined by Expert Council in Main Sanitary – Epidemiological Department of Health Ministry of KR in harmony with the Law "About safety of food-stuff" (2007). Normative documents for more effectiveness control of BAS were developed with help of Kazakh Academy of Nutrition. However, in KR there is no law about obligatory certification of this production. It is in our opinion not right.

Food-additives (preservatives, stabilizations, dyestuffs and others) are classes of substance with potential danger for human although there are the least dangerous pollutants. However, it is known that some preservatives, dyestuffs, enzymes could be the reason for developing allergy and other intolerance reactions. It could be for example sulfites (especially for patients with asthma), benzoic acid and parabenzenes (there are spasmolytics and rarely used like food additives, now they are applied in cosmetic production only). For improving control of food additives, it was developed "Technical Regulations by Controlling for Using of Food-Additives" in harmony with safety requirements of WHO and EC.

The pollution of food-products by radionuclide in KR due to Semipalatinsk nuclear testing area is a great importance. There was conducted a lot of over- and underground nuclear explosions, so now these territories present a great level of radionuclide pollution (Kirgizbaeva, 2006).

It is very important to study the ecological condition of this region because the use of some spaces in this region like pasture is debatable. The influence of radiation on the human and animal population through chain water–soil–plants–animals–food-products–human is not sufficiently documented in KR.

It was established that specific activity of radionuclide (Caesium-137, Strontium-90, and Plutonium-239) in the plant from explosion epicenters of Semipalatinsk region was respectively 52.2, 61.75 and 217.0 times higher of norm. For a distance from epicenters about 1–20 km this data were lower. Average activity of radionuclide in these zones was higher of norm respectively in 21.7, 10.8 and 1.1 times.

The development of new oilfield in Kazakhstan is increasing from year to year. Unfortunately pollutions of environment take place as a result of this process and of course food-products is contaminated by toxic compounds from oilfield. Therefore using of soils in oilfield regions must be accompanied by obligatory monitoring of pollutants cumulating in plants, especially in wheat, potatoes, cabbage, carrots, beetroot, berries and another agricultural production. The same situation occurs in chromium biochemical region of Ust-Kamenogorsk department and other region of Kazakhstan.

3. Recommandations for Food Safety

Thus, for successful solution of food-products quality and safety in Kazakh Republic the next tasks are needed to solve:

1. Organization of the common system of quality and safety control of food-products from "farm to table".
2. Organization of Centre for control of food-products safety and quality which must work in correlation with Centre Sanitary-Epidemiology Administration of Health Ministry of KR and other Government Department of RK.
3. Development of standards and measurement data for food-products according to the Law "About safety of food-stuff" (2007), legislation of European Parliament and *Codex Alimentarius*.
4. Analysis of Government and International Standards in the field of methods for detection of chemical and biological pollutants in food-products.
5. Step-by-step integration of laboratories by up-to-date equipment.
6. Step-by-step accreditation of laboratories according to international demands.
7. Introduction into the laboratory practice the methods of GMO detection.
8. Developing the methods for analyzing adulterated food-products.
9. Monitoring of quality and safety of food-products in Republic by using of Computer Centre.

References

Directive 2002/46/EC of the European Parliament and of the council of 10 June 2002 on the approximation of the laws of the Member States relating to food supplements (Text with EEA relevance)/Official Journal of the European Communities, 2002, L183/51–L183/57
Hadzhibaeva, I.F., 2006, Developing the ways of stabilization microbiological indexes of food-stuff, improving methods of control there safety. Abstract of dissertation at the competition of degree of candidate of biological science, Almaty, 26 pp
Kalamkarova, L.I., Mamonova, L.P., Bagriantseva, O.V., Baisrekin, B.S., Nazhmetdinova, A.Sh., Kuandukova, A.K., Hadzhibaeva, I.F., 2004, Fundamental directions of National Politic in the field of safety of food-stuff, *Health and Illness*, 5(35):25–30
Kirgizbaeva, A.A., 2006, Ecologo-hygienically estimation of quality of water and peculiar properties of radionuclide accumulation and migration in soil, plants at the territory of Semipalatinsk nuclear testing area, *Series of Natural and Technical Sciences*, 2:67–70
Nazhmetdinova, A.Sh., 2005, Scientific ground of regulations of some pesticides residues in environment, developing the methods of their identification and schemes of monitoring. Abstract of dissertation at the competition of degree of doctor of medical science, Almaty, 49 pp

ANALYTICAL MEASUREMENT AND LEVELS OF DIOXINS AND PCBs IN BIOLOGICAL SAMPLES

JEAN-FRANÇOIS FOCANT, GAUTHIER EPPE, EDWIN DE PAUW
Center for Analysis of Trace Residues (CART), Mass Spectrometry Laboratory, Organic and Biological Analytical Chemistry, University of Liège, Belgium

Abstract: Due to the high lipophilicity of dioxins are related compounds, more than 90% of the human exposure to dioxins is due to consumption of lipid-rich food such as meat; milk and dairies, as well as fish and derived products. The measurement of these trace level analytes requires a complex and tedious procedure including sample extraction, sample clean-up, and physico-chemical analysis after chromatographic separation. The determination of dioxins and related compounds requires a method that provides extremely high sensitivity and selectivity. High-resolution gas chromatography (HRGC) in combination with high-resolution mass spectrometry (HRMS) is the 'gold' standard method. In selected ion monitoring (SIM) mode, HRGC-HRMS (10,000 resolution) can achieve detection limit at the femtogram (10^{-15} g) level for most of the dioxin congeners. This paper explains the specificities and the pitfalls that could be met by analysts during the analytical development of this particular technique. Sample preparation, cleanup, selection of GC columns, specific aspects of double focusing sector instruments in SIM mode, high-resolution in mass, quantification by isotope dilution technique, quality assurance and quality control requirements and future perspectives are discussed and illustrated. A brief discussion of the bio-analytical approach is also provided to complete the analytical picture. Levels and trends for dioxins in food are discussed in the final part of the manuscript in the context of European regulation.

Keywords: High-resolution gas chromatography (HRGC), high-resolution mass spectrometry (HRMS), polychlorinated dibenzo-*p*-dioxin (PCDDs), polychlorinated dibenzofurans (PCDFs), polychlorinated biphenyls (PCBs), isotope Dilution (ID), chemical-activated luciferase gene expression (CALUX) assay

1. Introduction

Humans all over the world are exposed to chemicals during their life time. Among the thousands of existing anthropogenic compounds, some are persistent and remain in the environment for years once generated. The variation in measured levels mainly depends on the fact that some are (were) synthesized as industrial products although others are released accidentally or as by-products. Broad ranges of toxicities can be observed. The duality level-toxicity usually indicates if measurements of particular chemical or family of chemicals should be implemented. Polychlorinated dibenzo-*p*-dioxins (PCDDs), polychlorinated dibenzofurans (PCDFs), polychlorinated biphenyls (PCBs) are the persistent organic chemicals that are the most often measured in various types of matrices during food safety programs, environmental monitoring, and epidemiological studies. All together, they represent more than 400 individual molecules (congeners), which have to be separated from each other to ensure distinctive quantification of the target ones.

Due to their persistence, PCDDs, PCDFs and PCBs are part of the so called persistent organic pollutants (POPs) group of compounds that also include some chlorinated pesticides. Since they have a high lipophilicity and resist transformation, they bioaccumulate in animal and human adipose tissues. Consumption of food is considered as the major source of non-occupational human exposure to PCDD/Fs with foodstuffs from animal origin accounting for more than 90% of the human body burden, with meat, dairy, and fish products being the main contributors.

Mass spectrometry is known to be the method of choice for the analysis of contaminants at trace levels. It provides not only a very specific quantification but also ensures the unambiguous identification of target compounds. A key figure of merit in trace analysis is the absence of interferences. It was quite rapidly achieved by a combination of high-resolution and high-mass accuracy using double focusing magnetic sector instruments. The mass spectrometric instrumentation has considerably evolved during the last two decades with the apparition of robust quadrupoles, quadrupole ion traps, new time of flights and Fourier transform instruments as well as powerful hybrid instruments (Focant et al., 2005a). An alternative to high resolution has been introduced by quadrupole ion storage mass spectrometer in MS/MS mode. However, sectors instruments still find in that field, selected applications for which their performances are unmatched by any other techniques.

2. Dioxins, Furans, and PCBs

PCDDs, PCDFs, and PCBs are industrial substances that have been classified as Persistent Organic Pollutants (POPs) by the United Nations Environment Programme (UNEP) (UNEP, 2001). PCDD/PCDFs have never been produced deliberatively but are released as accidental by-products from combustion processes or industrial synthesis of other chlorinated chemicals. Although neither dioxins nor furans have ever had any commercial applications, PCBs have been heavily synthesized since 1930 for a variety of industrial uses such as, for example, dielectric fluids in transformers and capacitors. PCBs, which were supposed to be confined in industrial setting can however, like dioxins and furans, be virtually found in all global ecosystems. Dioxins and furans are planar tricyclic compounds that have similar chemical structures and properties (Figure 1).

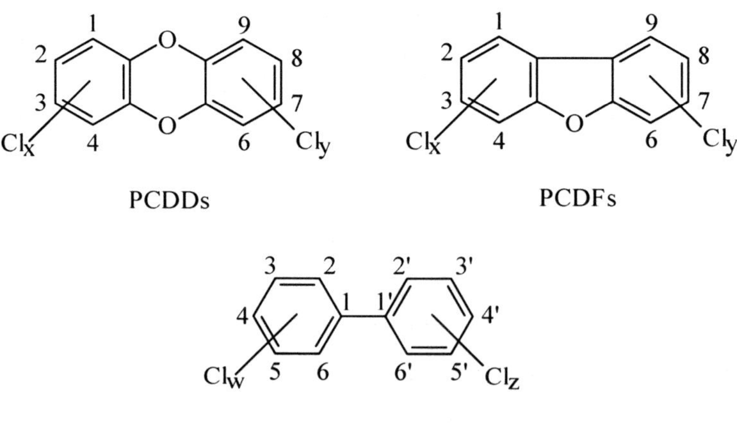

Figure 1. Chemical structure of PCDDs, PCDFs, and PCBs

PCBs are bicyclic compounds. The planarity depends on the chlorine substitution in positions 2,6,2′,6′ (*ortho* positions). Only non-*ortho*-chlorosubstituted PCB congeners adopt a planar geometry. These substitutional and geometrical parameters are of prime interest because they are closely related to the toxicity of these compounds. All together, dioxins (75), furans (135) and PCBs (209) represent 419 congeners that can theoretically be found in environmental samples and have to be separated for individual quantification (Table 1).

TABLE 1. Number of possible PCDD, PCDF and PCB congeners

Number of chlorine atoms	Number of isomers		
	PCDDs	PCDFs	PCBs
1	2	4	3
2	10	16	12
3	14	28	24
4	22	38	42
5	14	28	46
6	10	16	42
7	2	4	24
8	1	1	12
9	–	–	3
10	–	–	1
Number of congeners	75	135	209

Fortunately, most of them do not bio-accumulate in living organisms. Furthermore, for PCDDs and PCDFs, a 2,3,7,8-chlorosubstitution pattern is required for efficient binding to a specific cellular receptor responsible of toxic effects. This represents a group of 7 PCDDs, 10 PCDFs, to which 12 so-called 'dioxin-like' (DL) non-*ortho* and mono-*ortho* PCB congeners are added because of their structural similarity. A toxicity level has been assigned to each of those 29 compounds: the *toxic equivalency factor* (TEF). The TEF (Table 2) is a number that permits to translate the toxicity of a congener relatively to 2,3,7,8-TCDD, the most toxic congener (Van den Berg et al., 2006).

The concept of those relative toxicities permits to assess the total burden of a sample containing a mixture of PCDDs, PCDFs and DL-PCBs in terms of 2,3,7,8-TCDD concentration. The global toxicity or the *toxic equivalents* (TEQs) for a sample containing several different congeners is obtained using the formula:

$$TEQ = \sum_{n_1}(PCDD_i \times TEF_i) + \sum_{n_2}(PCDDF_i \times TEF_i) + \sum_{n_3}(PCB_i \times TEF_i) \qquad (1)$$

This translation of individual concentrations, issued from a physico-chemical analysis, in terms of TEQ facilitates risk assessment and regulatory control of levels and exposure to these compounds. Details concerning physical properties, formation, sources, toxicity, and structure-activity relationship can be found in the literature (Erickson, 1997).

TABLE 2. World Health Organization TEFs for humans

Congeners	WHO TEF	Congeners	IUPAC	WHO TEF
Dioxins		*Non-Ortho PCBs*		
2,3,7,8-TCDD	1	3,3',4,4'-TCB	77	0.0003
1,2,3,7,8-PeCDD	1	3,4,4',5-TCB	81	0.0001
1,2,3,4,7,8-HxCDD	0.1	3,3',4,4',5-PeCB	126	0.1
1,2,3,6,7,8-HxCDD	0.1	3,3',4,4',5,5'-HxCB	169	0.03
1,2,3,7,8,9-HxCDD	0.1			
1,2,3,4,6,7,8-HpCDD	0.01	*Mono-Ortho PCBs*		
OCDD	0.0003	2,3,3',4,4'-PeCB	105	0.00003
		2,3,4,4',5-PeCB	114	0.00003
Furans		3,3',4,4',5-PeCB	118	0.00003
2,3,7,8-TCDF	0.1	2,3',4,4',5-PeCB	123	0.00003
1,2,3,7,8-PeCDF	0.03	2,3,3',4,4',5-HxCB	156	0.00003
2,3,4,7,8-PeCDF	0.3	2,3,3',4,4',5'-HxCB	157	0.00003
1,2,3,4,7,8-HxCDF	0.1	2,3',4,4',5,5'-HxCB	167	0.00003
1,2,3,6,7,8-HxCDF	0.1	2,3,3',4,4',5,5'-HpCB	189	0.00003
1,2,3,7,8,9-HxCDF	0.1			
2,3,4,6,7,8-HxCDF	0,1			
1,2,3,4,6,7,8-HpCDF	0.01			
1,2,3,4,7,8,9-HpCDF	0.01			
OCDF	0.0003			

3. Analytical Aspects

3.1. SAMPLE PREPARATION

Environmental measurement of PCDD, PCDF and PCB occurs at the ultra-trace level (e.g. picogram or 10^{-12} g per gram of sample, parts-per-trillion – ppt). Such a low level can be illustrated by a situation where a drop of water would be present in a fleet of 1000 ultra large crude carrier super-tankers full of oil, which would make a line of ship of around 450 km long. Therefore, extremely large amounts of matrix-related inter- ferences have to be removed before one can even think about measurement. Extraction of the analytes from the matrix and purification of the target compounds from un- desirable interferences take place through an expensive and time consuming multistep approach. Although Soxhlet and liquid-liquid extraction (LLE) are still used for solid and fluid matrices, respectively, more recent and specific extraction methods exist. The major ones are supercritical fluid extraction (SFE), microwave-assisted extraction (MAE), pressurized liquid extraction (PLE), and solid-phase extraction (SPE). Several comparative studies and reviews are available in the literature for the user to elaborate an objective opinion on method characteristics (Camel, 2001).

Independently of the extraction method used, highly efficient clean-up procedures are required to purify samples issued from the extraction step prior the final analysis and quantification. Automated solid-liquid adsorption chromatographic separations, based on sorbents such as silica, alumina, Florisil, and activated carbon, are often used to ensure high sample throughput (Focant et al., 2004a). Another important aspect of the clean-up is the separation of the planar dioxins, furans and PCBs from the non-planar species. In practice, this fractionation results in a simplification of the gas chromatographic (GC) separation requirement prior to mass spectrometric (MS) analysis. Classically, a first fraction contains 17 PCDD/Fs and 4 non-*ortho*-PCBs, and a second fraction contains the eight mono-*ortho*-PCBs, as well as a group of six indicator PCBs (Aroclor 1260) that has also to be monitored because of their significance. The two fractions are subjected to GC-MS separation and analysis separately.

A strategy for efficient extraction, clean-up and fractionation rests on the use of a commercially available online automated integrated system. The system combines either the SPE or the PLE with multi-column clean-up. Figure 2 represents the plumbing diagram for such a system in the PLE version. In the PLE version, samples (meat, serum, …) are placed inside the extraction cell and extracted at elevated temperature and pressure. The eluant from extraction is directed to the clean-up columns where the fractionation also takes place. Separate fractions are collected and further evaporated prior GC-MS injection. For the SPE version, samples are directly loaded on the SPE column, dried under nitrogen flow and then eluted on the clean-up column. Table 3 and Figure 3 illustrate the performance of the integrated system versus other methods and certified reference materials.

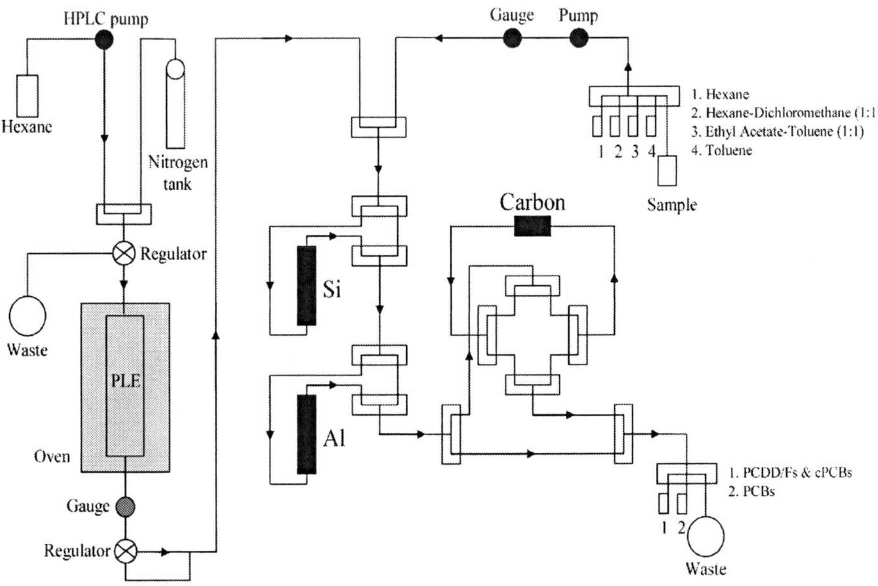

Figure 2. Plumbing diagram for the integrated PLE and clean-up system

TABLE 3. Repeatability and accuracy (acc) of Soxhlet, manual SPE and integrated SPE for the determination of PCDD/F concentrations (Cc) in certified reference milk powder material CRM 607 after reconstitution with water

Certified values			Soxhlet			Manuel SPE			Integrated SPE	
	Conc. (pg/g)	Cc. (pg/g)	RSD (%)	Acc (%)	Conc. (pg/g)	RSD (%)	Acc (%)	Conc. (pg/g)	RSD (%)	Acc (%)
2,3,7,8-TCDD	0.25 ± 0.04	0.29	3	116	0.27	13	108	0.34	6	134
1,2,3,7,8-PeCDD	0.79 ± 0.03	0.81	7	102	0.94	2	119	1.03	4	130
1,2,3,4,7,8-HXCDD	0.42 ± 0.07	0.46	8	110	0.47	15	113	0.40	10	95
1,2,3,6,7,8-HXCDD	0.98 ± 0.11	1.09	7	115	0.90	10	92	0.98	15	100
1,2,3,7,8,9-HXCDD	0.34 ± 0.05	0.38	8	113	0.33	3	98	0.37	8	110
2,3,4,7,8-PeCDF	1.81 ± 0.13	1.81	5	101	1.98	1	109	2.23	6	123
1,2,3,4,7,8-HXCDF	0.94 ± 0.04	0.92	8	101	0.86	15	91	0.87	8	93
1,2,3,6,7,8-HXCDF	1.01 ± 0.09	1.08	3	109	1.14	8	113	1.10	6	109
2,3,4,6,7,8-HXCDF	1.01 ± 0.05	1.07	8	100	1.18	2	110	1.18	8	110
Sum	**7.61 ± 0.61**	**7.92**	**4**	**103**	**8.08**	**2**	**105**	**8.49**	**4**	**110**

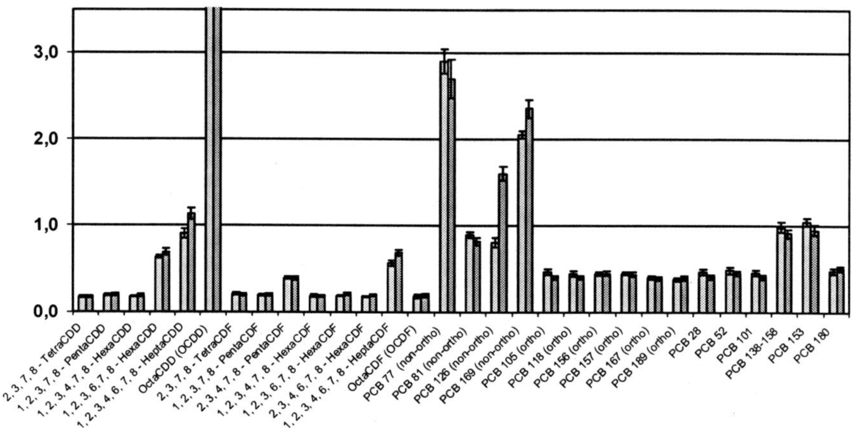

Figure 3. Data illustrating the efficiency (precision and accuracy) of the integrated PLE approach for quality control serum samples. Light grey is for the reference method, dark grey is for the integrated method (PCDD/Fs + cPCBs in ppt, other PCBs in ppb part per billion)

3.2. GAS CHROMATOGRAPHIC SEPARATION

Once extracted samples have been cleaned-up from matrix interferences and fractionated, the chromatographic separation of the target compounds has to take place before analysis. Due to the semi-volatility of dioxins and related compounds, gas chromatography (GC) is used to separate the different congeners and to allow non-ambiguous identifications. Capillary high-resolution GC (HRGC) columns ensure the required selectivity, especially for congeners of the same chlorination level. The separation characteristics and elution profiles on GC columns have been studied. An appropriate combination of column length, internal diameter and stationary phase polarity is needed.

For the analysis of human or animal-based food samples, non-polar GC columns are usually used. They allow separation between homologue groups and can also separate 2,3,7,8-substituted congeners (the ones that bioaccumulate) from each other (Liem, 1999). Usually 40–60 m columns with 0.18–0.25 mm internal diameter and 0.15–0.25 μm film thickness are selected. The 5% diphenyl/95% dimethyl polysiloxane stationary phase is a thermally stable (350°C) and low bleed stationary phase that conduct to a good separation of the 2,3,7,8 toxic congeners from others, especially for 2,3,7,8 TCDF for which 38 different substitution congeners are possible. The separation is however not always complete for hexachlorinated dioxin and furan congeners. Minimal separation requirements, as defined for example in the Commission Regulation No 1883/2006 (OJEC, 2006a), are nevertheless fulfilled and accurate measurement is possible. Planar non-*ortho*-PCBs, presents in the same fraction, are also separated using such a phase.

A little more polar stationary phase, the 8% phenyl polycarborane-siloxane, is often used for the separation of the PCBs contained in the second clean-up fraction. The carborane group has a high affinity for PCBs with a low degree of *ortho*-substitution. Although this phase does not allow the separation of all the 209 PCB congeners, it separates some critical pairs of co-elutions present with other phases (Larsen et al., 1995). In practice, the separation of the mono-*ortho*-PCBs and the indicator PCBs can be performed in 30 min on a 25 m × 0.25 mm × 0.25 μm column, a good compromise between the required resolving power and the GC run time. One should mention that to date, none of the existing stationary phases is capable of the separation of all the PCB congeners. Even emerging hyphenated methods such as comprehensive two-dimensional gas chromatography (GC × GC) coupled to time-of-flight (TOF)MS can at the most separate 192 congeners (Focant et al., 2004b). The GC × GC strategy consists in connecting 2 GC columns in series and to use a fast sampling device (the modulator) to create sharp injection pulses with the first column eluants into the second column. Coupling GC × GC to fast scanning MS such as TOFMS allows accurate description of the complex tri-dimensional chromatogram (Focant et al., 2005b).

3.3. HIGH-RESOLUTION MASS SPECTROMETRY

Mass Spectrometry is known to be the method of choice to quantify and identify trace level of organic compounds. Baughman and Meselson detected for the first time TCDD by HRMS in samples from Vietnam (Baughman and Meselson, 1973). But it is during the mid eighties and spurred on by Patterson and co-workers that sector instruments took on a new lease of life with dioxin applications (Patterson et al., 1986). The very particular nature of the analysis puts strong requirements on the detection method, leading to a less usual scanning technique, which is described hereafter.

Let us first summarize the specific aspects of the analysis having an impact on the mass spectrometric detection method. As mentioned above, the dioxins "family" is made of congeners corresponding to different chlorination levels (from 4 to 8 chlorine atoms for toxic ones). Within the same chlorination level, one has to deal with a different number of isomers of different toxicity, which must be specifically identified and quantified. To comply with regulations, the results of the analysis will reflect not only the quantity of individual congeners but also the global toxicity of the sample as described above in TEQ. Levels to be detected with confidence lay in the sub parts-per-trillion levels (ppt), leading to injected amounts in the picogram range down or even below to 0.1 pg for low contamination levels. Starting samples amounts can be high (e.g. several grams of fat for biological samples) which often will not allow more than two reinjections in case of technical problem. The method has to be very reliable.

Even after an extensive clean up and a high resolution chromatographic separation, the risk of interferences is still high. For that reason, the resolution ($M/\Delta M$) of the mass spectrometer should be set at least at 10,000 (10% valley definition). This allows mass discrimination at the 0.03–0.05 mass unit (dalton) level in the tetra- to octa-substituted congeners mass range.

To reach the required resolution, only double focusing sector instruments were historically available. Currently, the very high degree of analytical performances achieved by the HRGC-HRMS method is unmatched by any other techniques and Standards (EPA 1613, 1994; EN1948, 1996) require its use as the reference method for dioxin analysis. Fourier Transform Mass Spectrometry could compete in terms of resolution but with higher limits of detection and at higher cost. Time of flight mass spectrometer still have to compromise between resolution and acquisition speed, which can be useful for fast GC monitoring (Focant et al., 2004c). Bench-top low resolution quadrupole ion storage mass spectrometry is a useful alternative to HRMS (Plombey et al., 2000). The loss of mass resolution is counter-balanced by a gain in specificity in MS/MS mode.

With sector instruments, full scan spectra acquisition is not possible at such low levels. The instrumental limit of detection with that scanning mode is in the nanogram range. To gain in signal to noise ratio by a higher counting rate, the classical SIM method is used. The spectrometer is rapidly switched between the apexes of peaks of interest, avoiding wasting time and sample running in the chromatography while measuring between masses. Thus, the time spent (dwell time) on each target ions in the acquisition list is longer (in the 10–150 ms range, see Tables 4 and 5). The number of ions which can be measured at any one time is generally limited because at least ten sampling points for each GC peak are needed in order to get a Gaussian peak for accurate integration and quantification. Selected ions are therefore grouped in various segments. It is not the only reason of grouping ions in segments. Indeed, a high mass resolution means narrow peaks. A precise setting of the masses is mandatory. To control the mass accuracy, a lock mass is measured in each cycle and a lock mass check is often included (e.g. perfluorokerosene, PFK). The lock mass is ideally located within the measured mass range. A feed back loop compensated for any deviation in the mass setting.

TABLE 4. Target masses for mono-*ortho* PCBs in SIM mode for HRMS

	Window	Monitored ions		Ion dwell time	Interscan	Theoritica	15% for
	(min)	Quantitation ion	Confirmation ion	(ms)	(ms)	isotopic	isotopic
TCB $^{13}C_{12}$		303.9597[M + 2]	301.9626 [M]	26		0.77	0.65–0.88
PeCB		325.8804[M + 2]	327.8775[M + 4]	26		0.64	0.56–0.75
PeCB $^{13}C_{12}$	9–22	337.9207[M + 2]	339.9177[M + 4]	26	10	0.64	0.56–0.75
HxCB		359.8415[M + 4]	361.8385[M + 2]	26		0.81	0.69–0.94
HxCB $^{13}C_{12}$		371.8817[M + 4]	373.8788[M + 2]	26		0.81	0.69–0.94
Lock mass		316.9824 [I]	366.9792 [I]	1.3			
HpCB	22–28	393.8024[M + 2]	395.7994[M + 4]	96	10	1.04	0.88–1.20
HpCB $^{13}C_2$		405.8432[M + 2]	407.8402[M + 4]	96		1.04	0.88–1.20
Lock mass		366.9792 [I]	404.9760 [I]	4.1			

Syringe standard added prior to GC-HRMS analysis and used for recovery

TABLE 5. Target masses for PCDD/Fs and non-*ortho* PCBs in SIM mode for HRMS

Congeners	WHO TEF	Congeners	IUPAC	WHO TEF
Dioxins		*Non-Ortho PCBs*		
2,3,7,8-TCDD	1	3,3′,4,4′-TCB	77	0.0003
1,2,3,7,8-PeCDD	1	3,4,4′,5-TCB	81	0.0001
1,2,3,4,7,8-HxCDD	0.1	3,3′,4,4′,5-PeCB	126	0.1
1,2,3,6,7,8-HxCDD	0.1	3,3′,4,4′,5,5′-HxCB	169	0.03
1,2,3,7,8,9-HxCDD	0.1			
1,2,3,4,6,7,8-HpCDD	0.01	*Mono-Ortho PCBs*		
OCDD	0.0003	2,3,3′,4,4′-PeCB	105	0.00003
		2,3,4,4′,5-PeCB	114	0.00003
Furans		3,3′,4,4′,5-PeCB	118	0.00003
2,3,7,8-TCDF	0.1	2,3′,4,4′,5-PeCB	123	0.00003
1,2,3,7,8-PeCDF	0.03	2,3,3′,4,4′,5-HxCB	156	0.00003
2,3,4,7,8-PeCDF	0.3	2,3,3′,4,4′,5′-HxCB	157	0.00003
1,2,3,4,7,8-HxCDF	0.1	2,3′,4,4′,5,5′-HxCB	167	0.00003
1,2,3,6,7,8-HxCDF	0.1	2,3,3′,4,4′,5,5′-HpCB	189	0.00003
1,2,3,7,8,9-HxCDF	0.1			
2,3,4,6,7,8-HxCDF	0.1			
1,2,3,4,6,7,8-HpCDF	0.01			
1,2,3,4,7,8,9-HpCDF	0.01			
OCDF	0.0003			

The chromatographic challenge is then to bring the compound by groups (chromatographic windows) with no overlap, which would result in loss of congener measurement. Tables 4 and 5 give an overview of the HRMS acquisition parameters such as the time windows, the ions monitored, the dwell times and the isotopic ratios in SIM mode for non-ortho PCBs, PCDD/Fs and mono-ortho PCBs respectively. The sensitivity of the technique is considerably improved and HRMS manufacturers can guarantee a typical specification of 100 fg injected with a signal to noise ratio of 100:1 for TCDD at 321.8936 m/z in SIM mode with a mass resolution of 10,000 (10% valley definition).

3.4. IDENTIFICATION OF PCDDS, PCDFS AND PCBS

With such a scan, the identification of the compounds in terms of mass spectra is lost but can be confirmed by two other types of available information: the retention time (identification) and the measurement of the isotopic composition of the ions (identification and confirmation of the absence of interference). The presence of congeners affected by different TEFs leads to severe requirements for isomeric separation, which is relied on HRGC (EPA 1613, 1994, EN1948, 1996; OJEC, 2006a). The retention time of native and labelled standard peaks must be within a range of two seconds. To control the chlorination level and therefore the identity and the absence of interfering compounds, the measurement of the isotopic composition of the two most intense ions of both native and ^{13}C-labelled ion clusters must be ±15% around the theoretical ion abundance ratio (Tables 4 and 5). Any deviation out of this range will cause rejection of the congener's result. Figure 4 shows a typical chromatogram for TCDD. In practice, four ions traces (two native and two ^{13}C-labelled ions) per congener are plotted separately for integration and quantification. In the case of TCDD, an additional 1,2,3,4 TCCD (internal standard) is used to calculate recovery rates. This standard, also called the syringe standard, is added prior to HRGC/HRMS analysis. It enables to calculate the percentage of recovery standard (used for quantification) added prior to extraction. EPA (1613) recommends the use of the labelled $^{13}C_{12}$ 1,2,3,4 TCCD but other possibilities exist. In this example, $^{13}C_6$ 1,2,3,4 TCCD can be used to calculate recovery rates and to check the mass resolution at the same time (Patterson et al., 1989). For this purpose, the mass 331.9078 m/z ($^{13}C_6$ 1,2,3,4 TCCD [M+6]) is monitored and a resolution of 11,400 is required to completely resolve in mass 331.9368 m/z ($^{13}C_{12}$ 2,3,7,8 TCCD [M]) from 331.9078 m/z. By measuring the small peak that appears at the retention time of the recovery standard in the internal standard trace window (i.e. 331.9078 m/z) and by calculating the ratio with the peak 331.9368 m/z, one can check if the resolution is at least 10,000 during the GC run.

3.5. ISOTOPIC DILUTION TECHNIQUE FOR QUANTIFICATION BY HRMS

The power of mass spectrometry in quantitative analysis can further be enhanced by the isotopic dilution technique. This technique consists of spiking samples with an ideal recovery standard, which is an isotopically labelled standard (e.g. $^{13}C_{12}$ 2,3,7,8 TCDD), showing almost identical characteristics to the compound of interest (the native compound, e.g. $^{12}C_{12}$ 2,3,7,8 TCDD) during the extraction, the clean-up, the GC separation (almost the same retention time). The small mass difference (e.g. 12 m/z) enables the discrimination between the compound of interest and its recovery standard (Figure 4).

However, a discrepancy between native and labelled standards can be observed during electronic ionization (EI). A corrective factor has to be taken into account for

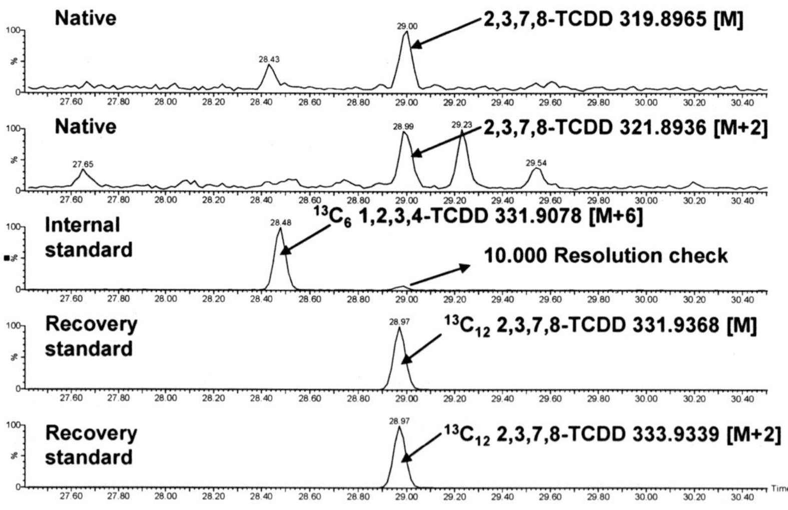

Figure 4. Isotopic dilution technique, SIM windows for TCDD including native, internal standard and recovery standard traces

accurate quantification. This factor is calculated with a minimum of five calibration points performed for all the PCDD/F and DL-PCB congeners with known amounts of native and recovery standards encompassed within the working range.

By plotting the concentration ratio of native to analogue ^{13}C-labelled against the area ratio of native to analogue standard ^{13}C-labelled, the slope of the calibration curve gives the corrective factor also called the relative response factor (RRF) (Figure 5). The RRF_i is calculated by the following equation:

$$RRF_i = \frac{(A^1_{native,i} + A^2_{native,i}) \times C_{l,i}}{(A^1_{labelled,i} + A^2_{labelled,i}) \times C_{n,i}} \tag{2}$$

Where $A^1_{native,i}$ and $A^2_{native,i}$ are the areas of the quantitation and confirmation ions for the native congener i, $A^1_{labelled,i}$ and $A^2_{labelled,i}$ are the areas of the quantitation and confirmation ions for its corresponding labelled compound i, $C_{n,i}$ is the concentration of the native compound i in the calibration solution and $C_{l,i}$ is the concentration of the labelled compound i in the calibration solution.

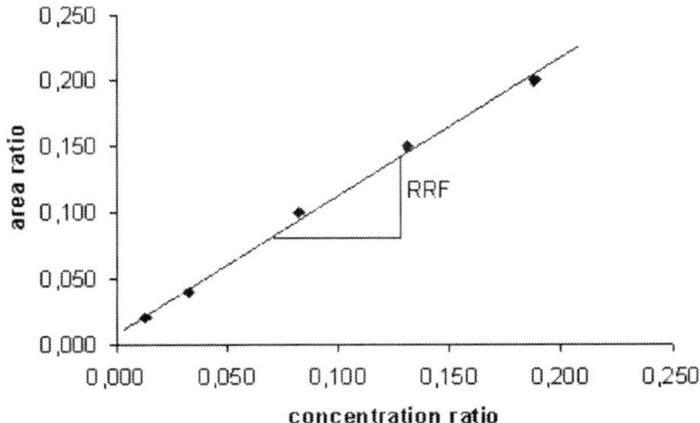

Figure 5. Five calibration points for 2,3,4,7,8 PeCDF

The mean RRF value over the five-point calibration range has to fulfill precision criteria according to the requirements of the Standards. Thus, the RRF value directly affects the congener quantification as indicated in the following equation:

$$[congener]_i = \frac{(A^1_{native,i} + A^2_{native,i}) \times Q_i}{(A^1_{labelled,i} + A^2_{labelled,i}) \times RRF_i \times m} \tag{3}$$

Where $[congener]_i$ is the concentration of the congener i (e.g. ng/kg), Areas are defined above (see equation 2), Q_i is the amount of the corresponding recovery standard i spiked (e.g. ng) in the sample , RRF_i is the relative response factor of the congener i and m is the weight of the sample (e.g. kg). Finally, the quantification expressed in TEQ is calculated according to equation 1.

3.6. QUALITY ASSURANCE AND QUALITY CONTROL (QA/QC)

The objective of QA/QC programs is to control analytical measurement errors at levels acceptable to the end-user of the data and to assure that the analytical results have a high probability of acceptable quality. It is generally evaluated on the basis of its measurement uncertainty associated with the final result which should match the end-user requirements. Accreditation according to the Standard ISO/CEN 17025 (ISO17025, 2005) is an adequate way to implement QA/QC programs in a routine laboratory. Dioxin analysis does not escape from the rule. We already mentioned some quality criteria for identification and quantification for those compounds by HRGC/HRMS. The above referenced standards and the IUPAC recommendations (Eurachem, 1998) provide guidelines and a comprehensive list of requirements that have to be fulfilled. We just want to mention some of them which are of primary importance when carrying out ultra trace level analysis.

One of the most important features of a QA/QC program is the use of procedure blank and internal quality control (IQC) samples. They have to be implemented with the series of real samples in the daily routine work. Both have to follow as much as possible the entire analytical procedure. IQC have to be characterized by sufficient homogeneity and long term stability to ensure that the analytical system is in control. When available, they should as much as possible match analyte, level and matrix of real samples tested. Control charts (plots of the data from blank or IQC vs time) provide the most effective mechanism for interpreting data. Blanks results and IQC data are both reported graphically on QC charts. These graphs provide the performances of the analytical system. Limits and decision rules have to be implemented in order to maintain the system

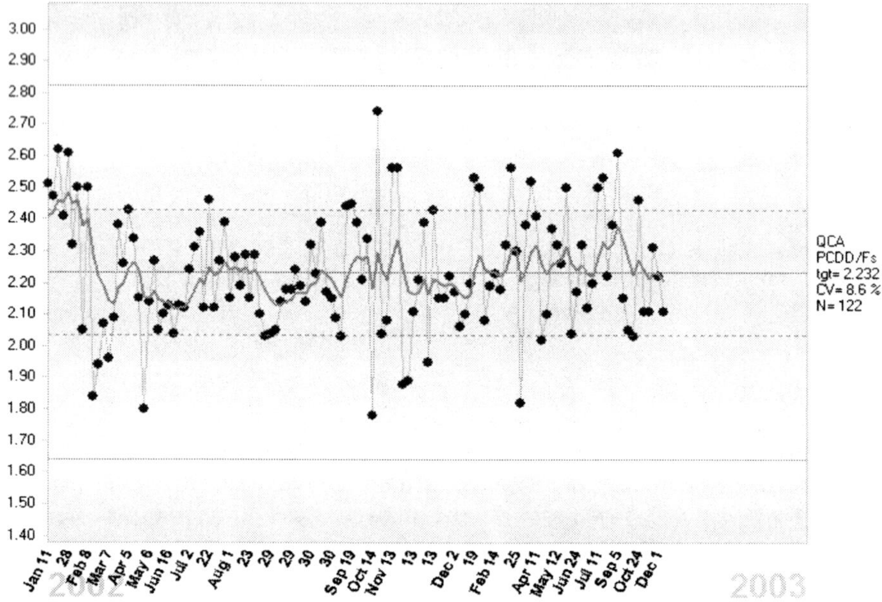

Figure 6. Quality control chart of PCDD/Fs expressed in TEQ in animal feedstuffs

stabilized and under control (Figure 6). The strategy of the decision process, the corrective and preventive actions to be taken when a lack of control is observed, should be clearly identified and followed.

The use of certified reference materials (CRMs) and the participation at relevant interlaboratory studies complete the requirements for assessing the accuracy of the analytical method (Eppe et al., 2004). Statistics and chemometrics are tools that are more and more involved in the extraction of the information from QA/QC data with the ultimate aim of improving the quality of the analytical results.

3.7. BIOLOGICAL ANALYTICAL APPROACH

There are two main groups of biological tools that can be used to respond or recognize contaminants such as dioxins: (1) bioassays that include cytochrome P450 induction assays, and (2) ligand binding assays that include enzyme immunoassays (EIAs) and receptor binding assays (RBAs). They are sometimes all unified under the bioanalytical detection methods (BDMs) appellation (Behnisch et al., 2001).

A well described RBA that is based on the ability of many dioxin-like planar aromatic hydrocarbons to interact with the AhR is the chemical-activated luciferase gene expression (CALUX) assay. It is an *in vitro* recombinant receptor/reporter gene assay system using several aspects of the AhR-dependent mechanism of action (Aarts et al., 1995). It is based on the use of the gene coding for firefly luciferase as a reporter gene. A plasmide containing luciferase gene under transcriptional control of DREs was made from mouse cell lines conducting to a recombinant mouse cell line showing AhR-controlled luciferase expression. The plasmide was further stably transferred to rat hepatoma cell lines, already containing the various factors involved in the AhR pathway, producing the final assay. Following ligand binding to the AhR, the heteromeric complex is transformed in its DNA binding form and binds to specific sites adjacent to the luciferase gene (Figure 7). The induction of the luciferase activity is dose-dependant; the intensity of the light emission is proportional to the concentration of AhR ligands and can usually be detected at lower levels than EROD. Furthermore, the potency of PCDD/F and PCB congeners to induce CALUX activity shown to be in accordance with TEFs with detection limits close to 1 pg. The applicability of the CALUX has been reported for various matrices like sediments, water, biological fluids, serum and milk.

Such biological methods however detect the overall exposure of many xenobiotic compounds via AhR mediation and can therefore not be used exclusively for diagnosis. If it is a good tool for comparison of global dioxin-like activity between groups, discrepancies arise between absolute values and those reported from GC-MS exxperiments. Even if it has proven to be robust enough to accommodate 'dirty' sample extracts, a dioxin specific clean-up is required to overcome the lack of specificity by removing non dioxin compounds and reduces over estimation risks regarding GC-MS TEQ values. Refinement of the methodology in order to separate activities due to various classes of chemicals is under investigation. Most of the proposed methodologies are however tedious and require a lot of sample handling. Such excessive manipulations are definitely time-consuming easily yield to poor recoveries and high risk of cross-contaminations.

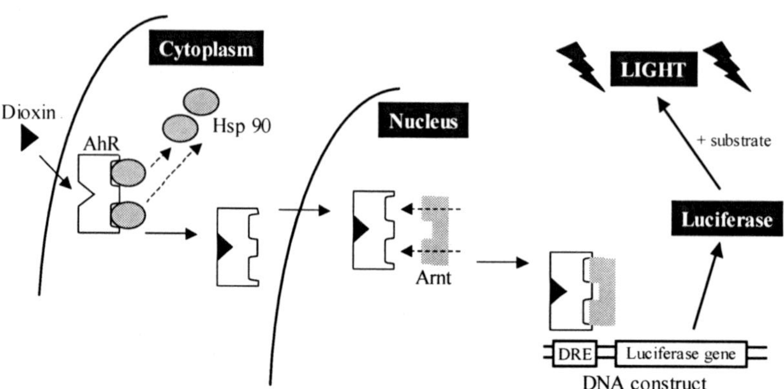

Figure 7. Simplified scheme of the CALUX-bioassay

The CALUX can nevertheless be used in the normative context if specific matrix references are used (Focant et al., 2005b). Figure 8 shows how the CALUX response compared to the reference GC-IDHRMS method. It was concluded from this comparison exercise that it is of prime importance to correct the raw CALUX data using a QC sample made of the exactly same matrix and characterized by similar congener distribution.

A cost estimate can be drawn based on the following: 1,000 unknown samples to be run yearly, including scientist employment, instrumentation purchase and paying-off (5 years), reagents, sample preparation, ID standards, consumables, technology licensing and royalties (DR-CALUX), and costs relating to the incorporation of the required amounts of blanks and quality control samples in the series of unknowns (Table 6). The cost estimate can vary, depending on several parameters, but the relative comparison of the methods presented here is based on similar operating conditions and parameters (DR-CALUX cost is based on duplicate sample measurement). The cost distribution is different. For the CALUX, if the cost contribution related to the measurement itself is reduced, the cost for scientific employment is increased. Using potentially more simple (cheaper) measurement technologies seems to result in higher human input requirements (data processing, reviewing, and reporting). Although, in this study, the MS-based technique used automated sample preparation steps and DR-CALUX used manual ones, the contribution of sample preparation to the global cost is similar. The use of the DR-CALUX approach does not offer congener-specific data and pattern description but permits the cutting of prices by half, making it an economically interesting screening method.

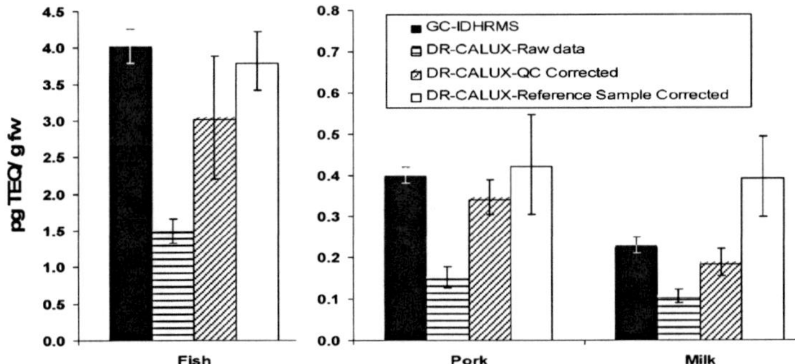

Figure 8. Raw and corrected responses of the DR-CALUX assay versus GC-IDHRMS for selected food samples

TABLE 6. Estimated percent distribution of the cost of the various stages of the measurement methods in the case of feed samples

	GC-IDHRMS	DR-CALUX[a]
Scientist employment	23	36
Extraction	11	7
Clean-up	28	29
Measurement	38	8
Licensing and royalties	–	20
Cost per sample (relative)	+++++	+++

[a]Cost based on duplicate measurements

4. Levels and Trends in Food

4.1. LEVELS IN FOODSTUFFS

Among food chain routes that have received particular attention, the dairy industry and fishery products are the major investigated ones. Data on concentrations of dioxins in foodstuffs are available for most EU Member States. Generally, data are adjusted on a lipid basis to normalize at each matrix lipid content. A report of a European Scientific Cooperation (EU SCOOP) based on results issued from more than 10 European countries is available since June 2000 (European Commission, 2000a). As illustrated in Figure 9, foodstuffs of animal origin present higher concentrations than those of plant origin.

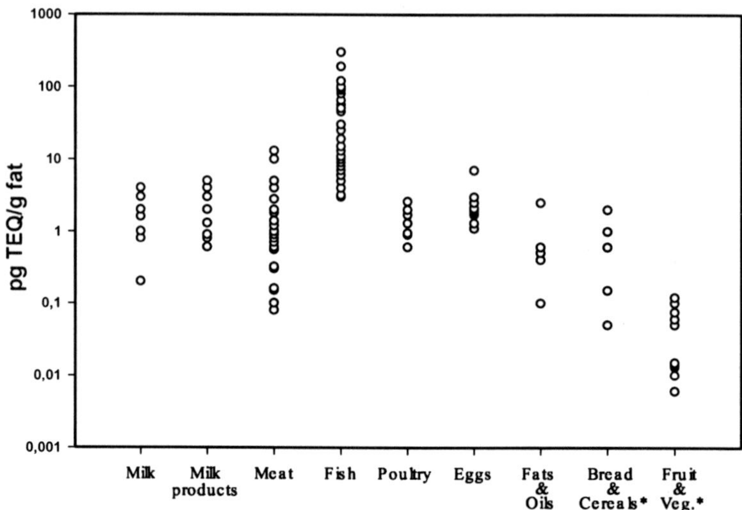

Figure 9. Concentration of dioxins in food (*in pg TEQ/g fresh weight) (European Commission, 2000a)

The largest range of concentration is observed for fish due to the wide difference in fat content and age of fish analyzed. Milk, milk products, meat, poultry and eggs have a mid point concentration in the range of 1–2 pg TEQ/g fat although a range of 2–214 pg TEQ/g fat is estimated for both wild fishes and farmed freshwater fishes. If wild fish's contaminations are due to bioaccumulation from surrounding environment, it appeared that farmed freshwater fishes contamination depends on the contribution of individual feed materials incorporated in their diet. A recent study pointed out that feedingstuffs for fish contained significant levels of dioxins, especially in Europe. European fish stocks are therefore more heavily contaminated than South Pacific stocks (SCAN, 2000). Great concern then arises when such fish-meals and oils are incorporated in diets of other food producing animals.

From the small set of data available for non-*ortho* PCBs, their estimated TEQ contribution varies of 0.5–1.8 pg TEQ/g fat for dairy products and of 0.2–2.4 pg TEQ/g fat for various type of meat (WHO, 1999). Average TEQ contribution of dioxin-like PCBs is usually between one to three times the TEQ contribution of PCDD/Fs (European Commission, 2000b). No clear pattern or geographical variation can be observed since there are not enough data points for statistical analysis. The only significant difference is observed for fish for which Baltic Sea specimens express higher levels. If mid point concentrations are calculated without Baltic Sea fish, the range decreases to 2–50 pg TEQ/g fat (European Commission, 2000a).

4.2. EU REGULATIONS ON FOOD AND FEED

Food safety issues, regarding to dioxin levels, are quite a new concern in most of European Union countries (OJEC, 2001). After many revisions, integration of human tolerable intake values and data on general background levels in EU, risk management conducted to the use of a three pillars strategy including maximum, action and target levels approach (OJEC, 2002).

Maximum levels: set as an exposure reduction measure (regulatory control), they must be below mean background contamination to be effective, they are strict but feasible levels.

Action levels: set using percentiles of the distribution of the contaminant levels for each food matrix, early warning tool for competent authorities. Any exceed of such levels should motivate efforts to identify sources and pathways of contamination. Measures should then be defined and applied to reduce the contamination. Such pro-active approach, based on a comprehensive set of measures, is expected to gradually allow reduction of dioxin levels in food and feedstuffs. Interactions with environmental measures are critical.

Target levels: lower than the current average background levels and could only be reached after further reduction of emissions. They indicate levels to be achieved over time in feed and food in order to bring human exposure for the majority of the EU population below the tolerable intake for PCDDs, PCDFs, and dioxin-like PCBs. They will be determined by 2008.

Table 7 presents some of the values concerning recommendations for regulation on food and animal feed, as available in Commission Regulation (EC) No 1881/2006 and Commission Directive 2006/13/EC (OJEC, 2006b, c).

In general, most of estimations of the dietary exposure are made by combining data concerning consumption habits and data on PCDD/F and dioxin-like PCB concentrations present in different types of foodstuffs. This approach is usually easier to carry out than duplicate meal collection. Since few years, dietary exposure is reported in terms dietary intake expressed in pg TEQ/day or in pg TEQ/kg b.w. (body weight)/day. In order to protect populations from intolerable exposure, studies have been conducted to produce a *tolerable daily intake* (TDI) value. Such a value results from the integration of exposure, toxicokinetic and health effects data. Since the first establishment of a TDI of 10 pg/kg b.w. for TCDD by the WHO in 1990, new epidemiological and toxicological data emerged. In 1998, experts revisited the TDI value based on the health risk for infants, cancer and no-cancer endpoints in humans and animals, mechanistic aspects, kinetic behavior, modeling, exposure, and applicability of the TEQ concept (van Leeuwen et al., 2000).

The lowest-observed-adverse-effect-level (LOAEL) for the most sensitive adverse responses reported in experimental animals were associated with body burdens from which a range of estimated long-term human daily intakes of 14–37 pg/kg/day was calculated. An uncertainty factor of ten was further applied in order to account for the use of LOAEL instead of no-observed-adverse-effect-level (NOAEL), to account for inter- and intra-species susceptibilities, and to account for differences in half-lives of complex TEQ mixture. The TDI was therefore finally expressed as a range of 1–4 pg TEQ/kg b.w. for dioxins, furans, and dioxin-like PCBs (WHO, 2000). More recently, the Scientific Committee on Food from the European Community proposed a *tolerable weekly intake* (TWI) value of 14 pg TEQ/kg b.w. (SCF, 2000). This value results from similar approach but based on slightly different studies. The weekly format was preferred by the Committee due to the very long half-lives of dioxins in the human body. Following the tendency to establish tolerable intakes on larger time-scale, a *tolerable monthly intake* (TMI) of 70 pg TEQ/kg b.w./month has also been proposed by the joint FAO/WHO Expert Committee on Food Additives (JECFA) (FAO/WHO, 2001).

Informations derived from food surveys in numerous of industrialized countries indicate a daily intake of PCDDs and PCDFs in the order of 50–200 pg TEQ/day or 1–3 pg TEQ/kg b.w./day for a 60 kg adult (WHO, 1999; SCF, 2000). If dioxin-like PCBs are also included, TEQ intake may be greater by a factor of 2–3. Regarding to food group contribution to the human dietary intake, it is generally accepted that meat and meat products, milk and dairy products, and fish and fish products account for roughly one third of the body burden, depending of countries (Liem et al., 2000).

TABLE 7. Examples of EU maximum levels in food products and in feedstuffs

	Maximum levels (pg WHO TEQ/g fat[a])	
	PCDD/Fs	PCDD/Fs + PCBs
Food products		
Meat and meat products from		
– Ruminants (bovine, sheep)	3.0	4.5
– Poultry and farmed game	2.0	4.0
– Pigs	1.0	1.5
Liver and derived products	6.0	12.0
Muscle meat of fish and fishery products[b]	4.0	8.0
Milk and milk products including butter fat	3.0	6.0
Hen eggs and egg products	3.0	6.0
Oils and fats		
– Animal fat		
– From ruminants	3.0	4.5
– From poultry and farmed game	2.0	4.0
– From pigs	1.0	1.5
– Mixed animal fat	2.0	3.0
– Vegetable oil	0.8	1.5
– Fish oil	2.0	10.0
Feedingstuffs		
Feed materials of plant origin	0.75	1.25
Vegetable oils and by-products	0.75	1.50
Minerals	1.0	1.50
Animal fat, including milk fat and egg fat	2.0	3.0
Other land animal products, including milk and milk products, and eggs and egg products	0.75	1.25
Fish oil	6.0	24.0
Fish, other marine animals, their products and by-products with the Minerals	1.25	4.5
Mineral additives	0.75	1.50
Feedingstuffs for fish and pets	2.25	7.0

[a]Upperbound values calculated assuming all non-detect equal to LOQ
[b]In pg PCDD/F TEQ/g fresh weight

Whatever tolerable intake approach is applied (daily, weekly or monthly), it appears that, even if levels are constantly decreasing, a situation free from concern is still not reached. Furthermore, a considerable proportion of the population still clearly exceeds the threshold values based on both PCDD/Fs and dioxin-like PCBs. Finally, since it is possible that repeated low-dose exposure to dioxins could substantially increase in body burden levels equivalent to a single, high-dose exposure, any efforts should be carried out to exit such situation.

4.3. BODY BURDEN

Body burden is defined as the total amount of a chemical in the body. Although the measure of the dose of dioxins in human was performed on adipose tissues in the early 1980s, requiring patient surgery, studies demonstrated that analyzing easily available serum samples correlated in an excellent way with adipose tissues in a three order of magnitude targeted dynamic range (Patterson et al., 1988). This was due to the partitioning of dioxins to various compartments in the body and use of lipid corrected values. Most of body burden estimations are, since then, carried out on serum samples.

Since the end of the 1980s, it is believed that the average dioxin concentration has been decreasing for 5–12% per year (European Commission, 2000a). The European population seems anyway to be more impregnated in average than other populations (Schecter et al., 1994). Range values for Europe (14–43 pg TEQ/g lipid) are higher than those reported for North America (19–27 pg TEQ/g lipid) and Japan (20–22 pg TEQ/g lipid) (Kumagai et al., 2000). Currently, it is generally accepted that (1) the annual increase in dioxin concentrations in the body is close to 0.3 pg TEQ/g fat (due to continuous ingestion of contaminated foodstuffs), (2) the dioxin and dioxin-like PCB body burden is around 20 pg TEQ/g lipid for international general population, and (3) efforts to reduce emissions are having beneficial effects. Furthermore, the use of pharmacokinetic model to predict trends in future exposures allows to state that current body burden could fall to below 10 pg TEQ/g fat by 2020 (Lorber, 2002). This is in accordance with trend data that show a constant decrease of dioxins and PCB levels in the environment since the 1970's.

At risk individuals can be defined as those consuming higher than average amounts of fatty food of animal origin, especially fatty fishes. Peoples living in potentially contaminated area such a near municipal solid waste incinerators or in area where extensive chlorinated pesticides have been used are also at higher risk of exposure.

4.4. INFANT EXPOSURE

The major route of exposure for infants in early years of life is consumption of contaminated breast milk, one of the most important sources of dioxins ever. Due to its relatively high lipid content and its human origin, the milk can contain fairly high levels of dioxins. Results obtained from second round of WHO coordinated assessment of dioxin concentration in breast milk of EU Member States in 1993 are presented in Table 8.

As expected, levels measured in industrial areas were always higher than in rural areas. One can also note that, by the time of the study, Belgium expressed one of the worst situations. If these results are however compared with a similar study carried out

TABLE 8. Average reported concentrations of dioxin in human breast milk

	Concentrations (pg TEQ/g fat)		
	Average	Minimum	Maximum
Rural	17.7	10.9 (Austria)	25.5 (Spain)
Urban	19.2	10.7 (Austria)	26.6 (Belgium)
Industrial	24	20.9 (Germany)	27.1 (Belgium)

5 years earlier, it appears that concentrations of dioxins in breast milk are in constant diminution at a rate of approximately 5–8% per year. Based on these levels and on average milk consumption, one can estimate infant exposure via consumption of mother's milk. Estimates then range from 100 pg TEQ/kg b.w./day in rural areas to 140 pg TEQ/kg b.w./day in industrial area. Since they have lower body weights and proportionally higher consumption rates during the nursing period, it is therefore generally observed that breast-fed infants present dioxin intakes of one to two orders of magnitude higher than adults, on a body weight basis. The situation is similar in other countries as reported in a recent study which estimated the mean daily intake of the U.S. population and shown a decline with age from 42 pg TEQ/kg b.w. for infants during the first year of life to 1.9 pg TEQ/kg b.w. at age of 80 years old and older. As a consequence of such intake, it has been reported that infants that were breast-fed for 4 months presented blood dioxin levels threefold higher than their mother at the age of 11 months and 13-fold higher than infants who were formula-fed (Abraham et al., 1996). It has further been estimated that, on average, the maternal body burden of dioxins decreases between at least 20 and 30% during lactation (Beck et al., 1994). Although situated in a critical period of neurological, physical and intellectual development, infants thus serve as dioxin exhauster for mothers which concentrate part of their body burden to a much smaller body.

Such case scenario is however quite the worst that can be set. One should take into account depuration rates of the milk itself that have been reported to be as high as a decline of 70% in concentration after 6 months of daily breast-feeding (LaKind et al., 2001). Furthermore, infants reduce their exposition quite rapidly as they gain weight and move to a mixed diet. In addition, the intake by breast-fed infants was mimicked in the studies that served to establish the TDI, in which the animal offspring was exposed through the suckling phase. Additionally, TDI is based on an average life-time exposure and it might be assumed that short high level exposure can be counter-balanced lower level of exposure in later life. Consequently, WHO promotion and support of breast feeding, which is associated with beneficial effects such as maternal bonding and protection against infections, should therefore still be encouraged (LaKind et al., 2000). In addition, subtle effects reported in studies were found to be associated with transplacental rather than lactational exposure.

Finally, infant exposure (peri- and post-natal) appeared to be determined by the life-time body burden of the mother before the conception. Short-term dietary measures before release or during breast feeding in order to reduce dioxin concentration in human milk would not succeed. Long-term reduction of the intake of these pollutants is required. Because human intake of dioxins is mainly due to consumption of animal fats, more attention must be paid to reduce exposure of animals. Regulatory efforts should therefore focus on identification and control of environmental sources, rather than recommendations concerning breast-feeding.

5. Conclusions

As illustrated above, the measurement of dioxins in biological matrices is extremely complex and tedious to perform. It requires high cost equipment and highly trained scientific staff. Because of the complexity of the matrices, one has to implement a multi-step procedure that can be divided in extraction, clean-up, and analysis. Some of the steps can be automated and coupled together to reduce the complexity and cost of the analysis. Because the final measurement step takes place at the low picogram level, strong QA/QC guidelines have to be respected to ensure proper analytical work and data integrity.

As the responses are given in toxic equivalents (TEQ), biological tests measuring the global toxicity are attractive solutions. However many compounds interfere with the biological activity or are not yet known. Large discrepancies may occur, which makes bioassays mostly useful as screening tests. Confirmation and random analysis using HRGC/HRMS remains mandatory. In addition, congener specific analysis brings additional useful information on patterns and sources of contamination.

The general trend to develop faster analytical methods is also present in the dioxins field. A perspective is the total integrated approach between sample preparation and GC/MS analysis. By coupling extraction technique, clean-up steps and GC/MS detection, an alternative to rapid screening bioassays is conceivable. In that case, the bottle neck of a congener-specific method will be the data processing and reporting.

References

Aarts, J.M., Denison, M.S., Cox, M.A., Schalk, A.C., Garrison, P.A., Tullis, K., Haan, L.H.J., Brouwer, A., 1995, Species-Specific Antagonism of Ah Receptor Action by 2,2′,5,5′-Tetrachloro- and 2,2′,3,3′,4,4′-Hexachlorobiphenyl, *Eur. J. Pharmacol. Environ. Toxicol.*, 293: 463–474.

Abraham, K., Knoll, A., Ende, M., Päpke, O., Helge, H., 1996, Intake, Fecal Excretion, and Body Burden of Polychlorinated Dibenzo-*p*-Dioxins and Dibenzofurans in Breast-Fed and Formula-Fed Infants, *Pediatr. Res.*, 40: 671–679.

Baughman, R., Meselson, M., 1973, An Analytical method for detecting TCDD (dioxin): levels of TCDD in samples from Vietnam, *Environ. Health Perspect.*, 5: 27–35.

Beck, H., Dross, A., Mathar, W., 1994, PCDD and PCDF exposure and levels in humans in Germany, *Environ. Health Perspect.*, 102(Suppl. 1): 173–185.

Behnisch, P.A., Hosoe, K., Sakai, S.-I., 2001, Bioanalytical Screening Methods for Dioxin and Dioxin-Like Compounds – A Review of Bioassay/Biomarker Technology, *Environ. Int.*, 27: 413–439.

Camel, V., 2001, Recent extraction techniques for solid matrices – supercritical fluid extraction, pressurized fluid extraction and microwave-assisted extraction: their potential and pitfall, *Analyst*, 126: 1182–1193.

EN1948, 1996, European Standard, Stationary source emission – determination of the mass concentration of PCDDs/PCDFs – Part 3: Identification et Quantification, December 1–51.

EPA method 1613 revision B, Tetra-through Octa-Chlorinated Dioxins and Furans by Isotope Dilution HRGC/HRMS, October 1994, 1–86.

Eppe, G., Cofino, W.P., De Pauw, E., 2004, Performances and limitations of the HRMS method for dioxins, furans and dioxin-like PCBs analysis in animal feedingstuffs: I. results of an inter-laboratory study, *Anal. Chim. Acta*, 519(issue 2): 231–242.

Erickson, M.D., 1997, *Analytical Chemistry of PCBs*. CRC, Boca Raton, FL.

Eurachem, 1998, The fitness for purpose of analytical methods, a laboratory guide to method validation and related topics, December 1998. Available at http://www.eurachem.org/guides/valid.pdf

European Commission, 2000a, Compilation of EU Dioxin Exposure and Health Data, Task 4 – Human Exposure, DG environment, United Kingdom Department of the Environment, Transport and the Regions (DETR). Available at http://europa.eu.int/comm/environment/dioxin

European Commission, 2000b, Opinion of the SCF on the Risk Assessment of Dioxins and Dioxin-Like PCBs in Food, Health and Consumer Protection Directorate-General, Directorate C – Scientific Health Opinions, Unit C3 – Management of Scientific Committees II, Scientific Co-operation and Network.

FAO/WHO (Food and Agriculture Organization of the United Nations/World Health Organization), 2001, Joint Expert Committee on Food Additives and Contaminants Meeting, Rome, Italy.

Focant, J.-F., Pirard, C., De Pauw, E. 2004a, Automated sample preparation-fractionation for the measurement of dioxins and related compounds in biological matrices: a review, *Talanta*, 63: 1101–1113.

Focant, J.-F., Sjödin, A., Patterson, D.G., 2004b, Improved separation of the 209 polychlorinated biphenyl congeners using comprehensive two-dimensional gas chromatography-time-of-flight mass spectrometry, *J. Chromatogr. A*, 1040: 227–238.

Focant, J.F., Cochran, J.W., Dimandja, J.M.D., De Pauw, E., Sjödin, A., Turner, W.E., Patterson, D.G., 2004c, High-throughput analysis of human serum for selected polychlorinated biphenyls (PCBs) by gas chromatography-isotope dilution time-of-flight mass spectrometry (GC-IDTOFMS), *The Analyst*, 129: 331–336.

Focant, J.-F., Pirard, C., Eppe, G., De Pauw, E., 2005a, Recent advances in mass spectrometric measurement of dioxins, *J. Chromatogr. A*, 1067: 265–275.

Focant, J.-F., Eppe, G., Scippo, M.-L., Massart, A.-C., Pirard, C., Maghuin-Rogister, G., De Pauw, E., 2005b, Comprehensive two-dimensionnal gas chromatography with isotope dilution time-of-flight mass spectrometry for the measurement of dioxins and polychlorinated biphenyls in foodstuffs. Comparison with other techniques, *J. Chromatogr. A*, 1086: 45–60.

ISO17025, 2005, The International Organization for Standardization, General Requirements for the Competence of Calibration and Testing Laboratories, September 2005.

Kumagai, S., Koda, T., Miyakita, H., Yamaguchi, K., Katagi, P., Yasuda, N., 2000, Polychlorinated dibenzo-p-dioxin and dibenzofuran concentrations in the serum samples of workers at continuously burning municipal waste incinerators in Japan, *Occupational and Environmental Medicine*, 57: 204–210.

LaKind, J.S., Berlin, C.M., Park, C.N., Naiman, D.Q., Gudka, N.J., 2000, Methodology for characterizing distributions of incremental body burdens of 2,3,7,8-TCDD and DDE from breast milk in North American nursing infants, *J. Toxicol. Environ. Health A*, 59: 605–639.

LaKind, J.S., Berlin, C.M., Naiman, D.Q., 2001, Infant exposure to chemicals in breast milk in the United States: What we need to learn from a breast milk monitoring program, *Environ. Health Perspect.*, 109: 75–88.

Larsen, B., Cont, M., Montanarella, L., Platzner, N., 1995, Enhanced selectivity in the analysis of Chlorobiphenyls on a carborane phenylmethylsiloxane copolymer gas-chromatography phase (HT-8), *J. Chromatogr. A*, 708: 115–129.

Liem, A.K.D., 1999, Basic Aspects of Methods for the determination of dioxins and PCBs in foodstuffs and human tissues, *Trends Anal. Chem.*, 18: 429–439.

Liem, A.K.D., Atuma, S., Becker, W., Darnerud, P.O., Hoogebrugge, R., Schreiber, G.A., 2000, Dietary Intake of dioxins and dioxin-like pcbs by the general population of ten European countries. Results of EU-SCOOP Task 3.2.5 (Dioxins), *Organohalogen Compd.*, 48: 13–16.

Lorber, M., 2002, A Pharmacokinetic model for estimating exposure of Americans to dioxin-like compounds in the past, present, and future, *Sci. Tot. Environ.*, 288: 81–95.

OJEC (Official Journal of the European Communities), 2001, Communication from the Commission to the Council, the European Parliement and the Economic and Social Committee, Community Strategy for Dioxins, Furans and Polychlorinated Biphenyls, 2001/C 322/02.

OJEC (Official Journal of the European Communities), 2002, EU Commission Recommendation (EC) No 2002/201/EC of March 4 2002 on the reduction of the presence of dioxins, furans, and PCBs in feedingstuffs and foodstuffs, 9.3.2002, L67/69–73.

OJEC (Official Journal of the European Communities), 2006a, EU Commission Regulation (EC) No 1883/2006 of December 19 2006 laying down methods of sampling and analysis for the official control of levels of dioxins and dioxin-like PCBs in certain foodstuffs, 20.12.2006, L364/32–43.

OJEC (Official Journal of the European Communities), 2006b, EU Commission Regulation (EC) No 1881/2006 of December 19 2006 setting maximum levels for certain contaminants in foodstuffs, 20.12.2006, L364/5–23.

OJEC (Official Journal of the European Communities), 2006c, EU Commission Directive 2006/13/EC of February 3 2006 amending annexes I and II to Directive 2002/32/EC of the European Parliament and of the Council on undesirable substances in animal feed as regards dioxins abd dioxin-like PCBs, 4.2.2006, L32/44–53.

Patterson, D.G., Holler, J.S., Lapeza, C.R., Alexander, L.R., Groce, D.F., O'Connor, R.C., Smith, S.J., Liddle, J.A., Needham, L.L., 1986, High-Resolution gas chromatography/high-resolution mass spectrometry analysis of human adipose tissue for 2,3,7,8-tetrachlorodibenzo-*p*-dioxin, *Anal. Chem.*, 58: 705–713.

Patterson Jr., D.G., Needham, L.L., Pirkle, J.L., Roberts, D.W., Bagby, J., Garrett, W.A., 1988. Correlation between serum and adipose tissue levels of 2,3,7,8-tetrachlorodibenzo-p-dioxin in 50 persons from Missouri, *Arch. Environ. Contam. Toxicol.*, 17, 139–143.

Patterson, D.G., Turner, W.E., Alexander, L.R., Isaacs, S., Needham, L.L., 1989, The analytical methodology and method performance of 2,3,7,8-TCDD in serum for the Vietnam veteran agent orange validation study, the ranch hand validation and half-life studies, and selected NIOSH worker studies, *Chemosphere*, 18: 875–882.

Plombey, J.B., Lausevic, M., March, R.E., 2000, Determination of dioxins/furans and PCBs by quadrupole ion-trap gas chromatography-mass spectrometry, *Mass Spectrom. Rev.*, 19: 305–365.

SCAN (Scientific Committee on Animal Nutrition), 2000, Dioxin Contamination of Feedinstuffs and Their Contribution to the Contamination of Food of Animal Origin. Available at http://europa.eu.int/comm/food/fs/sc/scan/out55_en.pdf

SCF (Scientific Committee on Food), 2000, Risk Assessment of Dioxins and Dioxin-Like PCBs in Food. Available at http://europa.eu.int/comm/food/fs/sc/scan/out78_en.pdf

Schecter, A., Fürst, P., Fürst, C., Päpke, O., Ball, M., Ryan, J.J., Cau, H.D., Dai, L.C., Quynh, H.T., Cuong, H.Q., Phuong, N.T.N., Phiet, P.H., Beim, A., Constable, J., Startin, J., Samedy, M., Seng, Y.K., 1994, Chlorinated dioxins and dibenzofurans in human tissue from general populations: a selective review, *Environ. Health Perspect.*, 102: 159–171.

UNEP (United Nations Environment Programme), 2001, Available at http://www.chem.unep.ch/pops/

Van den Berg, M., Birnbaum, L.S., Denison, M., De Vito, M., Farland, W., Feeley, M., Fiedler, H., Hakansson, H., Hanberg, A., Haws, L., Rose, M., Safe, S., Schrenk, D., Tohyama, C., Tritscher, A., Tuomisto, J., Tysklind, M., Walker, N., Peterson, R.E., 2006, The 2005 World Health Organization reevaluation of human and Mammalian toxic equivalency factors for dioxins and dioxin-like compounds, *Toxicol. Sci.*, 93: 223–241.

Van Leeuwen, F.X.R., Feeley, M., Schrenk, D., Larsen, J.C., Farland, W., Younes, M., 2000, Dioxins: WHO's tolerable daily intake revisited, *Chemosphere*, 40: 1095–1101.

WHO (World Health Organization), 1999, Joint FAO/WHO Food Standards Programme Codex Committee on Food Additives and Contaminants, Codex Alimentarius Commission, Food and Agriculture Organization of the United Nations.

WHO (World Health Organization), 2000, Consultation on Assessment of the Health Risk of Dioxins, Re-Evaluation of the Tolerable Daily Intake (TDI): Executive Summary, *Food Addit. Contam.*, 17: 223–240.

EXPIRED AND PROHIBITED PESTICIDES PROBLEM
IN UKRAINE

TAMARA GURZHIY
Independent Agency for Ecological Information, Kharkiv, Ukraine

Abstract: Twenty thousand to 25,000 t of expired or prohibited pesticides are stored on 4,000 Ukrainian depots. This is a serious threat for people and environment. Arsenic compounds are highly toxic for cattle. Death comes within several hours. Even if an animal managed to survive 7–10 days, it still may die because of kidney failure. There is no effective antidote. Organochlorine insecticides including DDT, lindane, toxaphene and heptachlor may be found in water food chain and achieve critical levels in fish-eating birds. Moreover, they are collected in cattle's fat. Poisoning by organochlorine insecticides affects nervous system and causes tremors, seizures and unusual behavior. After long-lasting seizures or high temperature animal dies because of respiratory failure.

Organophosphates and Carbamates insecticides are used for grain and seed treatment, cattle and pets. Young or exhausted animals are prone to be affected. Organophosphates are very different from each other by levels of toxicity. They enter organism through skin, lungs, eyes and digestive tract, and irritate nerves. Organophosphates, which contain chlorine atoms, usually leave fat slowly and cause long-lasting poisoning symptoms with pets and adult bulls. Carbamates act like organophosphates, but recovery period is usually faster. Severity of symptoms depends on pesticides concentration. The main antidote for such type of poisoning is atropine injection. It is difficult to cure ruminant animals because they keep a big volume of toxin in their stomach. Recovery takes days or weeks.

Sodium chlorate causes hemoglobin oxidation which leads to red blood cells death and destruction of muscles. Urine becomes brown. Cattle usually have diarrhea due to bowels irritation. There is no specific antidote. Forage treated with paraquat may irritate horse's mouth. If cats and dogs are exposed to that they may become fatally poisoned by licking their own feet. Majority of pesticide depots were not designed for long-term usage. Chemicals are stolen and illegally sold to people. Depots' roofs collapsed over the time, pesticides' wrapping gets of order, pesticides of different nature may become catalyst of spontaneous chemical reactions with unpredictable results. Spontaneous fire may spread toxins on a wide area. Utilization of expired and prohibited pesticides is Ukrainian national problem.

Keywords: Obsolete pesticides, organochlorine pesticides, nitrate fertilizers, carbamates

1. Introduction

Ukraine is famous for its fertile soil and well developed agricultural industry. Most natural resources are used for producing food, energy and other industrial goods. Simultaneously, Ukraine is presented as the most dangerously damaged territory due to Chernobyl Disaster and hazardous life-long pollution caused by different industries including heavy metal industry, defence industry etc. Approximately 70% of Ukraine's gross national income comes from the metallurgical, chemical and oil industries, which are the largest waste generators. Their wastes are deposited in public solid waste facilities. Most often industrial wastes are resulted because of usage of old-fashioned process technologies.

The human safety measures and ecological safety are directly related to the environment where people live. However, Ukraine has plenty of natural resources but most of them have been polluted due to uncontrollable industrial development. Ukraine now suffers from the effects of this industrial negligence. There is necessity for new technologies, modern manufacturing processes which should be based on clean technologies. The challenge is to utilize the resources of the natural environment in ways that protects public and occupational health without compromising the wellbeing of future generations. Many ecologically damaged regions in Ukraine are industrial areas with a high density of population. Seventy-two percent of surface is used for agricultural purposes. The most commonly acquired environmental problems are due to air pollution

B. Faye and Y. Sinyavskiy (eds.), *Impact of Pollution on Animal Products.*
© Springer Science + Business Media B.V. 2008

(86%), pollution of drinking water sources (83%), wrong way of waste recycling (81%), and some chemical pollution.

According to World Health Organization report, nowadays people are using 500,000 different chemical compounds during day-to-day activity. But more than 40,000 compounds have hazardous effects for environment and life organisms, and 12,000 are toxic. Most widely distributed pollutants are dust and ashes composed of oxides of ferrous and non-ferrous metals with different sulphur composite, nitric oxide, and some aerosols.

Ukraine generates an estimated 700–1,720 million tonnes of wastes per year. Low percent of solid waste is recycled. Mostly solid wastes are collected at surface dumps. Nearly, 5 billion tonnes are toxic wastes and less than 1% has been adequately recycled. As well-developed agricultural country, Ukraine was one of the first countries which were introduced for usage of pesticides. Deterioration of the humus layer of soils, due to poor management practices and old agricultural technologies, has lead to annual decrease in soil fertility, disruption of natural soil processes and loss in soil biodiversity. Decreasing of soil fertility is leading to excessive usage of fertilizers. According to the Institute of Soil Science and Agro-chemistry, 20% of Ukraine's land area is polluted (Kysil, 1999).

There are a lot of stocks of obsolete pesticides which remain in the country from 'Soviet Union time'. (Golubovska-Onisimova et al., 2006) Pesticides are stored in poor storage facilities. There is no government policy or some necessary precautions measures about obsolete pesticides. Very often obsolete pesticides are uncontrollably used in current agricultural process. This is required some government actions for safely storage of obsolete pesticides to prevent further leakage and spreading of hazardous substance. Pesticides must be removed in order to make new working conditions safe according to safe measures. Banned pesticides that have been left in stores from 'Soviet Union time' are leaking on to the floors of stores. Some pesticides produce vapour, some eventually turn into liquid which leak into the soil or some vaporized pesticides start dribbling on brick walls to the soil.

Pesticides with expired date of usage and other agricultural chemicals are stored at 109 sites under local government administration and at 4,000 sites at agricultural enterprises. Most of them should not be used due to ambiguous quality. Many of the storage sites pose environmental risks because of the high toxicity of the expired pesticides compounds and disturbed storage conditions. Eighty percent of storage sited does not take preventative measures to protect the atmosphere and soil form toxic pollution. Most pesticide stocks are located within city-boundaries and, that is why people are in constant threat for environment.

As at 1 December 2003, estimated stockpiles of OPs in the country reached 20,900 t (earlier, these pesticides were delivered to agricultural facilities, but had not been used) (Golubovska-Onisimova et al., 2006).

Open storages are the most common available storages. Pesticides in such storages are under direct sunlight which quickly change quality of pesticides and destruct them. Usually, there are no security or precautions measures to protect people from hazardous influence of expired pesticides. Usually, people have access to pesticides storages and take them to use it on their own private lots. Also, it was noticed that empty pesticide drums are often considered functional for domestic usage and often are sold in street market (Krainov, 1999).

The problem of obsolete pesticide recycling is serious and very urgent. Severe damage to human health from obsolete pesticides usage is expected. The importance of this problem is far complicated than it could be imagined. The lack of assistance and support, as well as lack of expertise is making that problem more urgent and immense. There are urgent needs for the guidelines, guidance and advices from other countries and some international donor sponsoring organizations.

2. Influence of Pesticides on Animals

2.1. NITRATE FERTILIZERS

In agricultural premises, especially during drought, excessive using of nitrate fertilizers is leading to excessive content in plants (Kysil, 1999). Plants have tendency to accumulate those fertilizers. Eventually after harvesting those plants are used for fodder. It is also possible that cattle get poisoning from sources of drinking water, mostly from superficial wells and from condensing water running from ventilation and irrigation canals. Therefore, accumulated nitrate fertilizers are getting to digestive tract of cattle where they are transforming into more toxic nitrites.

Signs of poisoning are depending on type, dose, time of exposure, physiologic condition of cattle and individual cattle sensitivity. Exposure might be acute (in case of taken of excessive amount of poisoned food) and chronic.

Signs of nitrate fertilizers poisoning are lost of appetite, flabbiness, eventual weight loss, as well as shortness of breath and pain. First signs of nitrate fertilizers poisoning are depending on dose of exposure. Usually, digestive system is damaged first, followed by respiratory distress and nervous system alteration including behavioural changes, lethargy, as well as some haematological changes.

2.2. ARSENIC POISONING

Arsenic compounds are highly toxic for cattle. Arsenic compounds can be detected in hairs of cattle. Arsenic poisoning leads to destruction of blood vessels followed by tissue necrosis. Digestive tract membranes get inflamed first. Acute arsenic poisoning causes hemorrhagic gastroenteritis. Most common first sign of toxicity is diarrhoea with abdominal pain, followed by weakness and exhaustion. Due to weakness affected cattle is often found near source of water. Death comes within several hours due to high central nervous system toxicity. Even if an animal managed to survive 7–10 days, it still may die because of kidney failure. There is no effective antidote. In order to confirm the fact of poisoning it is necessary to take samples of blood, urine, and hair.

2.3. ORGANOCHLORINE INSECTICIDES

Insecticides are very often used in agricultural premises and, therefore, they could be potentially determined as hazardous substances for livestock (Report NGO MAMA-86, 2000).

Organochlorine insecticides are chemically divided on two groups: compounds of aliphatic row and compounds of aromatic row. All pesticides are barely solved in water and have specific smells. Organochlorine insecticides include DDT, lindane, toxaphene, and heptachlor. Among fumigants (dichloretan, chlorpikrin, and paradichlorbensol) the most toxic is chlorpikrin which was used during WWI as warfare agent leading to respiratory failure.

Remains of organochlorine insecticides can accumulate in the fat of cattle until toxic levels are reached. Accumulated organochlorine insecticides have central nervous system toxicity which will lead to insidious onset of dyspnoea, muscle tremors and twitches, seizures, and unusual behaviour, preceded by high body temperature. Animals become comatose which followed by death due to respiratory failure.

Nowadays, lindane and endosulfan have labels for uses in the garden. Endosulfan is a significant risk to cats, cattle and dogs due to its acute toxicity. The most toxic is aldrin (isodrin, endrin). Toxic dose of aldrin is up to 5 mcg/kg in young cattle and 20–25 mcg/kg in adult cattle. If the dose of aldrin exceeds 25 mcg/kg in chronic exposure of young cattle, it will stop growth. In the amount exceeding 100 mcg/kg organochlorine poisoning signs can be detected (Shedey, 2001).

2.4. ORGANOPHOSPHATES AND CARBAMATE INSECTICIDES

Organochlorine insecticides include DDT, lindane, toxaphene and heptachlor. They may be found in water food chain and achieve critical levels in fish-eating birds. Mostly these compounds get to cattle organisms through skin, mouth, lungs and eyes. Malathion is one of the least toxic with terbufos (Counter), coumaphos (Co-ral), parathion, methyl parathion, guthion and di-syston being highly toxic.

Mechanism of action of Organophosphates and Carbamate Insecticides is inhibitory action for cholinesterase. These substances are more lipid-soluble with greater CNS access. Absorption of organochlorine derivatives from intestines occurs slowly. Most of the chemicals are removed from cattle organisms with faeces. However, DDT derivates in rabbits are removed with urine in the acetylated forms and only insignificant quantities of DDT can be found in bile. In contrast, cats DDT derivates do not accumulate.

Chronic intoxications are observed from lipid-soluble substances (DDT and Hexochloran). Among DDT, derivate Methoxichlor is removed quickly from organism and that is why there is no possibility for chronic intoxications. Starved and week animals with thinner fat layer are more sensitive to insecticides as well as younger animals, especially calves who get poisoning with milk from cows which were fed with insecticides treated forage.

Sensitivity in ascending order to organochlorine derivates: mice, a cat, a dog, the rabbit, a porpoise, the monkey, a pig, a horse, large horned livestock, a sheep and a goat. The fish is more sensitive to DDT, and birds, on the contrary, are less sensitive (Turusov et al., 2002). The main antidote for such type of poisoning is atropine injection. It is difficult to cure ruminant animals because they keep a big volume of toxin in their stomach. Recovery takes days or weeks.

Sodium chlorate causes haemoglobin oxidation which leads to red blood cells death and destruction of muscles. Urine becomes brown due to hematuria. Cattle usually have diarrhoea due to irritation of mucous membranes of digestive system. There is no specific antidote.

Forage treated with paraquat may irritate horse's mouth and stimulate excessive saliva secretion. Animals start to have uncontrollable chewing movement and become irritated. If cats and dogs are exposed to that they may become fatally poisoned by licking their own feet. Paraquat damages kidney, liver and lungs (PANNA, 2003). Cause of death is usually due to multiple organ failure.

3. Possible Solutions and Recommendations for Obsolete Pesticides Problem in Ukraine

- To strengthen environmental protection governance at the national, regional and local levels, including licensing of agricultural premises
- To increase accountability for planning and regulation by national and local agencies
- To involve citizens in the development of environmental policies and business plans as well as increase population's participation in major decisions concerning the use of natural resources
- To raise public awareness of environmental issues
- To establish incentives for the adoption of best available technologies for greater efficiency and the reduction of pollution
- To strengthen the legal framework and institutional capacities for solid waste disposal and management
- To enhance collaboration and coordination in pesticide disposal
- To prepare a target and funding plan for disposal operations
- To offer the possibility of exchanging information and experiences regarding recently completed disposal operations and of discussing collaboration in future disposal operations

References

Golubovska-Onisimova, A., Tsygulyova, O., Tsvetkova, A., 2006, Attention – Obsolete Pesticides! The problem of obsolete pesticides management in Ukraine, MAMA-86 All-Ukrainian NGO, Valentyna Shchokina, MAMA-86-Nizhyn, WECF Munich/Utrecht, 22 pp.

Krainov, I.P., Borovoi, I.A., Skorobogatov, V.M., 1999, Elimination of Obsolete Pesticides. Eco-technology and Resource Conservation, pp. 47–55.

Kysil, V.I., 1999, Application of fertilizers in biological agriculture, *Agrochemistry*, 10: 69–78.

Report of NGO activities in connection with the process of development of the Plan for reduction of risks of accumulated stockpiles of obsolete pesticides in Ukraine, rKyiv. All-Ukraine NGO MAMA-86, 2000.

Shedey, L.O., 2001, Influence of various systems of agriculture on parameters of microbiological activity of chrnozem typical, *Agrochem. Soil Sci.*, 61: 76–83.

Turusov, V., Rakitsky, V., Tomatis, L., 2002, Dichlorodiphenyltrichloroethane (DDT): Ubiquity, Persistence, and Risks, *Environ. Health Perspect.*, 110: 125–128.

Squibb K., Applied Toxicology: Pesticides at http://aquaticpath.umd.edu/appliedtox/pesticides.pdf

Pesticide Action Network North America (PANNA), 2003, Pesticide Action Network (PAN) International on Paraquat at http://www.panap.net/uploads/media/PAN_Intl_PP_Paraquat_Dec03.pdf

ASSESSING THE HAZARDS OF RADIOLOGICAL AND ENVIRONMENTAL FACTORS FOR THE PUBLIC HEALTH IN THE WESTERN KAZAKHSTAN

USEN KENESARIYEV, ZHAMILYA BEKMAGAMBETOVA, NIYAZALY ZHAKASHOV, YERZHAN SULTANALIYEV, MEIRAM AMRIN

S.D. Asfendiyarov Kazakh National Medical University, Chair of Common Hygiene and Ecology, Almaty

Abstract: The subject of this study was the pollution of the major natural environments in the area of nuclear explosion at the Azgyr nuclear test base. It has been identified that the soil of the nuclear sites, mineshaft water, bottom slits in reservoirs, wild flora, potatoes, cow milk are polluted with technogenic radionuclides. Since the territory of the nuclear test base was used for the disposal of various technical wastes, different heavy metals have been detected in those natural environments. The study of the health status in the population of the testing site area allowed identifying the group of disease classes in which morbidity rates are significantly higher than those of the reference inhabited locality. The correlation between the pollution of natural environments and the morbidity of people in the nuclear test base area has been established.

Keywords: Nuclear test base, radioactive nuclides, heavy metals

1. Introduction

In connection with a more detailed study of the problem related to the nuclear weapon tests or accidents at nuclear facilities, it became obvious that the increasing number of sick population was found in the zones affected by background ionizing radiation doses. Thus, the consequences of the Chernobyl accident (Rachinsky, 1991), nuclear tests at the Semipalatinsk nuclear test base (Ibrayev and Gusev, 1995) were well-known.

However, it should be noted that the so-called "peaceful nuclear explosions" (PNEs) were set off within the territory of CIS countries, with Kazakhstan having the second largest number of such explosions after Russia (Vassilenko, 2004). Altogether, 39 PNEs were set off in our country, 17 of which – having the yield of 1.1–103 kt – at the Azgyr nuclear test base which is located in the West of Kazakhstan, in Atyrau Region. This fact impelled to study radio-ecological situation in that region and to assess the hazard of man-triggered factors for the level of health of the people dwelling in the vicinity of the test base under study.

2. Materials and Methods

To study the problem, we made a hygienic assessment of quality of major natural environments (water from water sources, drinking water from wells, soil, vegetation, and some foodstuff items) in the area of the Azgyr test base, both within the territory of nuclear sites and nearby inhabited localities of Kurmangazy District, Atyrau Region (settlements of Balkuduck, Azgyr, Assan, Konyrtereck, Suyunduck, Batyrbeck, Zhalgyzapan, Ushtagan). Check tests were made in the inhabited locality of the same district, Ganyushkino settlement, located 200 km from the test base.

The overall population in the region under study was over 22,000, the indigenous population comprising the main population contingent, the majority of which (90.9–98.3%), according to the survey, was living in the region under study over 10 years. In the total number of population, male individuals were 50.4%, and female – 49.6%, with the proportion of children of 29.3%.

It is worth mentioning that at the time of the study the nuclear sites were not fenced and were easily accessible both for people and for domestic animals (mainly cows). In addition, we have found out that some of the cavities which appeared after the explosions had been used for the disposal of radioactive waste, over a long period of time.

The environmental samples were analyzed for radioactive nuclides and heavy metals. Specific activity of the majority of natural as well as technogenic radionuclides (^{137}Cs, ^{90}Sr) in the environmental samples was determined in the laboratory of the radiation department of the Republican Sanitary and Epidemiological Station (SES). The following methods were used for this purpose: radiometric, radiochemical, spectrometric and photocolorimetric tests that were made in several steps. High-sensitivity scintillation γ- and β-spectrometric PC-based system was used with the detector on NaI, Tl crystals with the specialized "Progress" software allowing decoding complex imaging spectra with low energy resolution.

Priority heavy metals in water, soil and other natural environments were detected with the help of atomic absorption method on the AAS-1 spectrophotometer. As the whole, we have analyzed the following number of surface water samples: for heavy metals – 286, inorganic compounds – 125, radionuclides – 120, and made over 4500 element determinations, underground water samples (from wells) – 105, 75 and 85, respectively (over 2,500 element determinations), water samples from decentralized sources (mineshafts) – 350, 175 and 150 (about 5,000 element determinations). Also, we analyzed bottom slits from reservoirs (over 600 element determinations), over 500 soil samples and about 200 samples of wild flora, potatoes, and cow milk.

3. Results

3.1. WATER CONTAMINATION

Tests of water and bottom slits from surface reservoirs for the presence of heavy metals showed that in the water sources of the test base area, iron concentrations ranged from 1.2 MCL (maximum concentration limit) to 4.3 MCL, lead – up to 3 MCL, and they were within the MCL in the control sample. The bottom slits from reservoirs in the test base area, as compared to the control sample, show the excessive contents of manganese (by 37 times), cobalt (by 12 times) and lead (by 240 times).

The quality analysis of drinking water from the mineshafts for the presence of heavy metals showed that the content of cadmium in a number of samples amounted to 3–9.8 MCL, and lead was present in drinking water of all inhabited localities in the area of the Azgyr test base, its content exceeding the permissible value: from 3.3 MCL (Azgyr settlement – 1.5 km from the nuclear site A-1) to 6.3 MCL (Suyunduck settlement – 40 km south west of A-2). These heavy metals were either absent in the well water of the reference settlement or their concentrations were much lower than the MCL.

Test for radioactive contamination of water from reservoirs and drinking water from decentralized sources in the Azgyr test base area showed that based on the overall alpha-activity water contamination in the reservoirs near the nuclear sites exceeded the "intervention level" (IL) and amounted to 34.6 IL, and based on β-activity – to 2.33 IL. Alongside with that, radionuclide composition of water from reservoirs in terms of natural radionuclides (^{226}Ra, ^{232}Th) was within permissible activity levels (PALs), and in terms of ^{90}Sr it exceeded the permissible value by 4.2.

It was determined at the same time that the activity of bottom slits with α-transmitters ranged between 130 and 1.5×103 Ku, and with β-transmitters – from 3.3×103 Ku to 3.7×104 Ku. The analysis of radionuclide composition of bottom slits from the lake located close to the A-9 nuclear site has shown that they contained ^{137}Cs – 3.0 ± 0.01 Bq/kg and ^{60}Co – 10.0 ± 0.01 Bq/kg, whereas these were not found in the reference reservoir.

The excess of the limits on aggregate indicators (α- and β-activity) by 1.3–141 times was found in the drinking water of the mineshafts in the Azgyr test base area whereas in the control sample they were within the limits. These indicators, compared to the control sample, showed the excess of 7–141 and 1.25–16.25 times, respectively.

The analysis of the quality of water in the wells for the presence of technogenic nuclides revealed that if ^{137}Cs and ^{90}Sr in the settlements near the Azgyr test base were found in quantities much less than the PAL (0.04 Bq/l or less), as well as at the sensitivity threshold of the method, they were not found in the drinking water of Ganyushkino settlement. However, bottom slits from mineshafts in the Azgyr test base area contained significant concentrations of ^{137}Cs (8.0 ± 0.01 Bq/kg – 18.0 ± 0.02 Bq/kg) and ^{60}Co (4.0 ± 0.01 Bq/kg –7.0 ± 0.01 Bq/kg) and that is not safe in terms of quality of drinking water in those sources.

3.2. SOIL CONTAMINATION

The assessment of soil quality in connection with technogenic pollution showed that across all nuclear sites the maximum concentration limits in the soil were exceeded as follows: copper – 9.7–18.7 MCL, lead – 2.2–3.6 MCL and cobalt (at A-1) – 3 MCL. According to the analysis, the content of other heavy metals (Mn, Pb, Cd) in the soil was within the concentration limits, however the background content of lead showed the excess of 1.4–1.5 times at all sites. In all examined inhabited localities of the test base area the average concentration of copper was 7 times higher than MCL, of lead – by 3.5 times, of cobalt – by 14.5 times and of cadmium – by 2.9 times. The lead content in the soils of inhabited localities close to the test base was within the established limits but exceeded the background value by 1.6 times on average. In the reference settlement of Ganyushkino the content of the above elements in the soil was within the limits.

Given that until now the natural radiation background was exceeded by up to 5.5 times within nuclear sites of the test base, we tested the soil samples for their radionuclide composition.

It was determined that the ground contamination with ^{137}Cs across all sites under study was not even and was of the local nature. Alongside with that the largest concentrations of that element were detected at the sites A-1 and A-5 – $4,083 \pm 30$ and $6,948 \pm 46$ Bq/kg, respectively. The lowest content of this radionuclide was found in the soil of A-10 site – from <2.8 to 759 ± 15 Bq/kg. The content of ^{90}Sr in the soil of nuclear sites ranged between <10 and 785 ± 100 Bq/kg. As for the group of naturally occurring radioisotopes, the largest concentrations of ^{40}K – from 15 ± 3 to 847 ± 32 Bq/kg – were detected in the soil of all nuclear sites of the test base. The surface soil layer in the settlements of the Azgyr test base area was found to contain technogenic ^{137}Cs (from <11 to 37 Bq/kg), as well as radioactive elements of natural origin – ^{40}K, ^{232}Th and ^{226}Ra. In the soil of the reference settlement of Ganyushkino technogenic ^{137}Cs was detected in the concentrations which were 1.4–4.7 times less than in the settlements located close to nuclear sites and that seemed to be caused by global radioactive pollution only.

3.3. FOOD PRODUCTS CONTAMINATION

It was obvious that contamination of drinking water as well as soil in the Azgyr test base area with technogenic radionuclides and heavy metals of various hazard classes should have various adverse effects on a human body. Their direct and indirect impact on a human body through drinking water and foods of plant and animal origin may be conductive to deterioration in health among the inhabitants of settlements near the test base.

We discovered that the content of a number of heavy metals in potatoes exceeded concentration limits: for lead – up to 5.8 PL (permissible level), for cadmium – up to 20 PL, the excess of maximum permissible level (MPL) have been found out for cadmium – 2 MPL, cobalt – 3.7–10.9 MPL, iron – 1.3 MPL. Specific activity of ^{137}Cs in potatoes grown in the test base area was within the permissible level, however, it was 3.9–5.8 times greater than in the check sample (P < 0,01) and on average 16.1 times less than in the wild flora.

The analysis of wild flora in the Azgyr test base area showed high coefficients of concentrations of a number of heavy metals: zinc –2.1 MPL, copper – 1.3 MPL, lead – 8.4 PL, cadmium – 6 PL, 2 MPL, cobalt – 12.3 MPL. Also, statistically reliable high concentrations of zinc, lead, cadmium, and cobalt (P < 0.05–<0.001) were found in the cow milk sample taken in the test base area, compared to the check sample, while the contents of ^{137}Cs and ^{90}Sr were within the PL and made up from 34.9 ± 1 to 54.5 ± 2.1 Bq/l and from 9.6 ± 3.2 to 12.7 ± 1.1 Bq/l, respectively.

3.4. HUMAN CONTAMINATION

The study of the health status among the population of the Azgyr test base area based on the in-depth medical examinations showed high general morbidity rates: from 3,131.8‰ (Azgyr settlement) to 4,256.8‰ (Balkuduck settlement), which exceeded the reference rates 1.5–3.5 times. At that, average morbidity rates of the whole population in the Azgyr test base area based on six disease classes (gastrointestinal diseases, diseases of urogenital system, circulatory system, diseases of blood and blood-forming organs, neoplasms, mental disorders and behavioral disorders) occupying the top places in the general morbidity structure, are 1.49–3.46 times higher than those of the reference Ganyushkino settlement.

The use of the multiple correlation and regression analysis between the population morbidity rates and the specific content of toxic elements in the drinking water or soil of that region allowed, for example, establishing dependence of the neoplasm Class morbidity among the population of the region under study and ^{137}Cs and Pb found in the soil and F, Cd, Pb, and ^{137}Cs (R > 0.97, P < 0.05) found in the drinking water.

4. Discussion and Conclusion

Long-term operation of the Azgyr nuclear test base resulted in the adverse radio-ecological situation, contamination of soil, flora, foodstuff (potatoes, milk), water bodies with technogenic radionuclides and heavy metals, thus currently presenting health hazard for the people dwelling nearby the test base. Hence, a package of structural, sanitary and hygienic, medical and preventive actions intended for environmental protection and reduction of the population's morbidity rates needs to be implemented.

References

Ibrayev, N.S., Gusev, B.I., 1995, Regarding the environmental condition in the Semipalatinsk Region, Healthcare of Kazakhstan, chapter 7: 39–40.
Rachinsky, V.V., 1991, Dose dependence of biological impact of radiation is the scientific foundation of radiation safety. Social After-Effects of the Chernobyl Disaster, In "The Thesis and the Speech at the Scientific Conference Based on the Launching of the Chernobyl" – Socium Program, Moscow, 17–18 December 1991, Agricultural Academy, Moscow, pp. 48–51.
Vassilenko, O.I., 2004. Radiation ecology, Medicina, Moscow, p. 215.

PART II

Relationships between soil contamination, plant contamination and animal products status

ORGANIC POLLUTANTS IN ANIMAL PRODUCTS

CHRIS COLLINS
*Department of Soil Science, School of Human and Environmental Sciences,
University of Reading, Whiteknights, Reading, UK*

Abstract: It is necessary to assess the contamination of animal products by pollutants in order to protect human health. This can be achieved by understanding the processes whereby contaminants are introduced into the food chain. Such contamination can arise from the atmospheric deposition on to crops and soil or via contaminated feed. These processes are described and quantified with particular reference to the deposition of organic pollutants onto pasture. The transfer of the contaminants from the fodder and feed to the animal is also described. Finally, these processes are put in context by the illustration of how they are used in regulatory exposure models.

Keywords: Pollutants in pasture, modelling, pollutant transfer, animal food contamination

1. Introduction

In order to understand the fate and transport of organic pollutants in the environment it is essential to have an understanding of the physical and chemical properties of the pollutants of concern. Two of the most widely used properties for the prediction of the behaviour of organic pollutants are the Henry's law constant and the octanol-water partition coefficient. The Henry's law constant describes the partition of the chemical under consideration between the air and the water phases and its derivation is given in equation 1:

$$H = P/C_{water} \tag{1}$$

where
H = Henry's law constant (Pa m^3 mol.$^{-1}$)
P = pollutant partial pressure in air (Pa)
C_{water} = pollutant concentration in water (mol. m^{-3})

This is often quoted as the dimensionless Henry's law constant or air to water partition coefficient (H' or K_{AW}).

$$H' = H/RT = K_{AW} = C_{air}/C_{water} \tag{2}$$

where
H = Henry's law constant (Pa m^3 mol.$^{-1}$)
R = the universal gas constant (Pa m^3 mol.$^{-1}$ K^{-1})
T = Absolute temperature (K)
C_{air} = pollutant concentration in air (must be same units as C_{water})
C_{water} = pollutant concentration in water (must be same units as C_{air})

The octanol-water partition coefficient describes the transfer between water and octanol. Octanol is used because it is a good surrogate for body lipids and a clean laboratory reference can be obtained for measurements. The derivation of the octanol-water partition coefficient is given in equation 3, because so many organic pollutants are so hydrophobic it is often expressed as the log K_{OW}.

B. Faye and Y. Sinyavskiy (eds.), *Impact of Pollution on Animal Products.*
© Springer Science + Business Media B.V. 2008

$$K_{OW} = C_{octanol}/C_{water} \qquad (3)$$

where
K_{OW} = octanol-water partition coefficient
C_{air} = pollutant concentration in air (mol m^{-3})
C_{water} = pollutant concentration in water (mol m^{-3})

One of the characteristics of many of the persistent organic pollutants is their high Kow value, which results in these compounds being retained in the fat tissues and transferred to milk as a consequence of their lipophilicity. They are also resistant to breakdown and bioaccumulative as a consequence of this property. Table 1 provides the H and K_{OW} values for the US EPAs 'dirty dozen' these are compounds of concern because of their toxicity and ability to accumulate in the food chain as identified by the Stockholm Convention.

TABLE 1. Physico-chemical properties of the US EPA 'dirty dozen'

Compound	Log K_{ow}	H (Pa m^3 mol^{-1})
Aldrin[a]	5.52	17.0
Chlordane[a]	5.54	48.6
DDT[a]	6.36	0.81
Dieldrin[a]	4.55	1.51
Endrin[a]	4.56	0.75
Heptachlor[a]	4.27	29.4
Hexachlorobenzene[a–c]	5.31	132
Mirex[a]	6.9	828
Toxaphene[a]	4.68	0.06
Polychlorinated biphenyls (PCBs)[b,c]	3.90–8.26	3.1–272
Polychlorinated dibenzo-p-dioxins[c]	4.30–8.20	1.22–22.7
Polychlorinated dibenzo-p-furans[c]	4.31–8.00	0.34–19.1

[a]Pesticide
[b]Industrial chemical
[c]By-product

The number of potential contaminants from industrial processes is considerably wider than the EPA 12, for example the Environment Agency of England and Wales has identified 27 possible organic contaminants from industrial land (DEFRA, 2002) these have been related to different industrial activities (Anonymous, 2006). Some examples are shown in Table 2.

There are a number of pathways by which these industrial chemicals can potentially contaminate animal products (Figure 1.) By investigating the pollutant profiles in food samples the principle source can be identified for example an Egyptian study found that solid waste burning was the most important source in the contamination of food and feed samples (Loutfy et al., 2007).

TABLE 2. Contaminants associated with industrial processes

Industry	Phenol	Acetone	Chloro-phenols	Oil/fuel hydrocarbons	Aromatic hydrocarbon	PAHs	Chlorinated aliphatic hydrocarbon
Airports		X		X	X		X
Animal and animal products works	X				X	X	X
Asbestos manufacturing works					X	X	X
Ceramics, cement and asphalt manufacturing works		X		X		X	
Charcoal works		X			X		X
Chemical works: coatings (paint and printing inks) manufacturing works	X				X	X	X
Chemical works: cosmetics and toiletries manufacturing works		X			X	X	X
Chemical works: disinfectants manufacturing works	X		X		X	X	
Chemical works: explosives, propellants and pyrotechnics manufacturing works	X	X		X	X		X
Chemical works: fertiliser manufacturing works				X		X	
Chemical works: fine chemicals manufacturing works	X	X			X	X	
Chemical works: inorganic chemicals manufacturing works						X	
Chemical works: linoleum, vinyl and bitumes-based floor covering manufacturing works	X			X	X	X	X

The two main pathways are ingestion of contaminated vegetation and contaminated food supplements. The inhalation of pollutant vapours is not considered to be important accounting for <5% of the pollutant load in the animal (McLachlan, 1996).

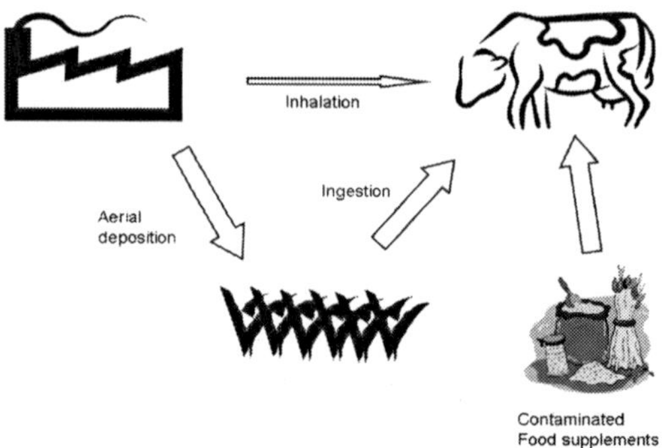

Figure 1. Pathways of organic contaminants into livestock products. Arrows indicate magnitude of pathway

2. Transfer from Atmosphere to Vegetation

Vegetation can be contaminated by a number of pathways: uptake from soil, deposition of gases and particles from the atmosphere, deposition of resuspended soil particles, volatilisation from the soil with subsequent aerial deposition. Gaseous uptake from ambient air has been demonstrated to be the main uptake pathway into plant foliage for a variety of organic chemicals, including PAHs (Simonich and Hites, 1994), PCBs (Bohme et al., 1999; Meneses et al., 2002) and tetra and hexa-chlorinated PCDD/Fs (Welsch-Pausch et al., 1995). A detailed description of these processes can be found in Collins et al. (2006). A brief description of the models used to describe these processes is provided here.

Simple relationships have been derived for the prediction of the plant concentration from the air concentration (Bacci et al., 1990), but a more comprehensive framework has been proposed by McLachlan (1999) which is derived from the octanol/air partition coefficient (K_{OA}) of the chemical under consideration, where $K_{OA} = K_{OW}/K_{AW}$ (Figure 2). Within this framework chemicals of high volatility but low lipophilicity have an equilibrium between air and plant concentrations, while the deposition of those of higher lipophilicity are considered to be kinetically limited and those of the highest K_{OA} are deposited with particulate matter. The kinetic limitation arises because the plant has a larger pool for uptake of the chemical than can be satisfied by the aerial concentration.

Briggs et al. (1982) derived a relationship for predicting concentrations of chemicals in the transpiration stream of the plant from the concentration in soil solution and the K_{OW} of the chemical (Figure 3). This relationship was based on experiments investigating the uptake of a limited number of non-ionized chemicals into barley plants. It was found that the transpiration stream concentration factor (TSCF) concentration in xylem/concentration in external solution) was at a maximum for chemicals with log K_{OW} ca. 1.8. This relationship has been widely used in both plant modelling (Trapp and Matties, 1995; Ryan et al., 1988) and also by some authors to explain their results (Groom et al., 2002; Hsu et al., 1990; Burken and Schnoor, 1998).

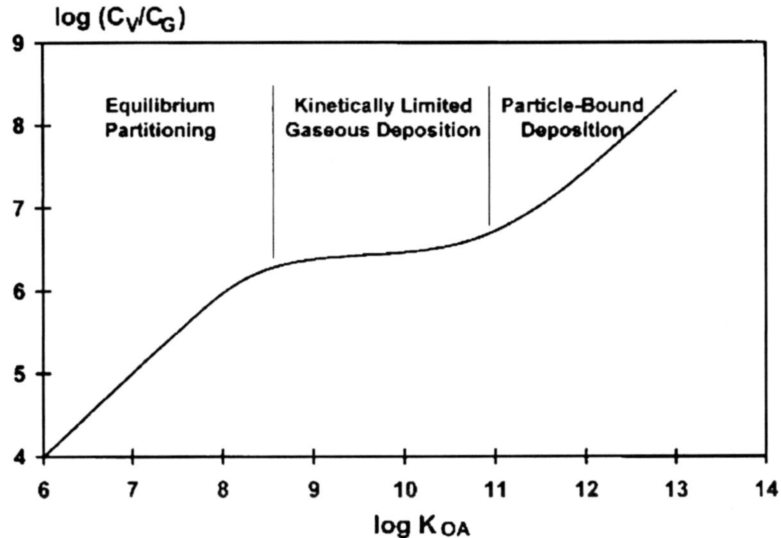

Figure 2. Interpretive framework of McLachlan (1996) illustrating the relationship between log KOA and uptake process for organic chemicals in the atmosphere (C_V = vegetation concentration, C_G = gas concentration, K_{OA} = octanol air partition coefficient)

The stem concentration factor (SCF = concentration in shoot/concentration in soil) which derives from this relationship is influenced by the fraction of organic carbon in the soil. There is a significant decline in the transport to the stem as the soil carbon increases and becomes a larger sink for the organic pollutants.

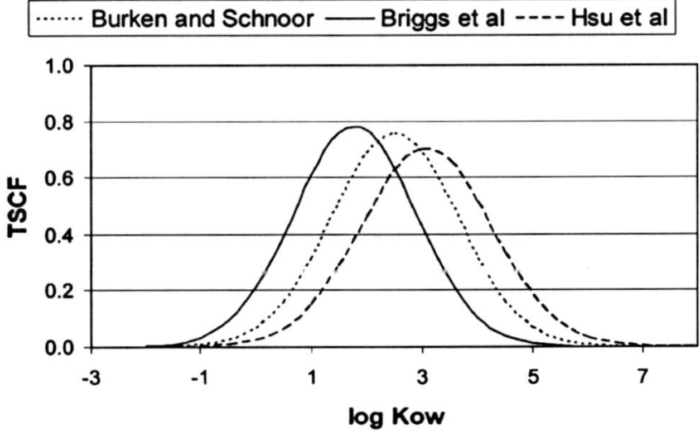

Figure 3. Variation of transpiration stream concentration factor with KOW after Briggs (1982)

3. Transfer from Vegetation to Livestock

Ingestion is the primary exposure route of organic pollutants to animals with very little by inhalation. This can be from contaminated pasture (Mamontova et al., 2007; Kierkegaard et al., 2007; Welsch-Pausch and McLachlan, 1998) or contaminated feed (Bernard et al., 2002; Traag et al., 2006; Costera et al., 2006). Pasture can be contaminated from deposition of POPs from the atmosphere to both the vegetation and soil. The latter is important as direct soil ingestion was reported to be the most significant pathway for contamination of milk, particularly in arid regions (Mamontova et al., 2007). Contamination of feed can be a consequence of its ingredients, particularly if this contains fish meal which can contain high levels of POPs because they are concentrated in the fish oils which are used in the formulation of the feed. This arises because POPs bioaccumulate in food chains and fish have a high lipid content to store these chemicals (Shaw et al., 2006). Accidental contamination of feed can also occur. One of the most famous cases was the contamination of animal feed by PCBs and dioxins in Belgium in 1999, up to 2,500 farms were potentially affected and some chicken meat contained 100 times greater than the regulatory limit (Bernard et al., 2002).

Once ingested significant quantities of the pollutants can be absorbed this is usually expressed by a carryover rate which details the amount of pollutant passing into the animal product of concern, e.g. milk or meat. Thus carryover rates are proportional to the absorption into the animal.

$$COR = \frac{m * fy}{f * F} * 100 \qquad (4)$$

COR = carry over rate (%)
m = pollutant concentration in fat at steady state (ng kg^{-1})
fy = fat yield (g day^{-1})
f = pollutant concentration in diet (ng kg^{-1} DM)
F = is daily feed intake (g day^{-1})

In a study with goats dioxins and furans had higher carryover rates with decreasing chlorination suggestion reduced absorption and increased metabolism of the higher chlorinated compounds (Costera et al., 2006), Similar findings have been reported by other workers (McLachlan, 1993; McLachlan and Richter, 1998). These have resulted in simple equations to predict the COR which can be used for modelling.

$$COR = -12\log K_{OW} - 108 \qquad (5)$$

$$C_M = (9q_{gr} + 4.5q_{cs}) C_A COR/q_f \qquad (6)$$

where
C_M = content in milk fat (mol m^{-3} fat)
q_{gr} = daily ration of grass (g d.w. day^{-1})
q_{cs} = daily ration of silage (g d.w. day^{-1})
C_A = atmospheric concentration (mol m^{-3})
q_f = rate of milk fat excretion (g day^{-1})

Equations such as these can form the basis of predictive models for the contamination of animal products. However, some caution needs to be used in their implementation. For example different classes of POPs have been observed to accumulate in different animal products milk for PCBs (Mamontova et al., 2007), meat for PBDEs (Kierkegaard et al., 2007). Additionally log K_{OW} may not be the only predictor of carryover rate, as the metabolism of compounds of similar K_{OW} has been observed to vary (Costera et al., 2006).

4. Modelling Exposure from Contaminated Animal Products

In order to determine the human exposure from contaminated products a number of models have been developed. One example is the Xtrafood model produced for the Belgian Science Policy Secretariat (Seuntjens et al., 2006). The model calculates transfer through the food chain using a series of equations such as those outlined above. Once these concentrations are known they are coupled to historical records of human food consumption data. The estimated human exposure from the total accumulated via ingestion of the individual foodstuffs is then combined for the overall exposure assessment. Models such as these can be used to assess risks from routine industrial discharges, the impact of spike releases as a consequence of industrial accidents or predict the impact of future installations.

References

Anonymous, 2006, Environment Dot "Industry profiles for 47 different industrial activities," P.O. Box. 29, The Stationery Office.

Bacci, E., Calamari, D., Gaggi, C., Vighi, M., 1990, Bioconcentration of organic-chemical vapors in plant-leaves – experimental measurements and correlation, *Environ. Sci. Technol.*, 24: 885–889.

Bernard, A., Broeckaert, F., De Poorter, G., De Cock, A., Hermans, C., Saegerman, C., Houins, G., 2002, The Belgian PCB/dioxin incident: analysis of the food chain contamination and health risk evaluation, *Environ. Res.*, 88: 1–18.

Bohme, F., Welsch-Pausch, K., McLachlan, M.S., 1999, Uptake of airborne semivolatile organic compounds in agricultural plants: Field measurements of interspecies variability, *Environ. Sci. Technol.*, 33: 1805–1813.

Briggs, G.G., Bromilow, R.H., Evans, A.A., 1982, Relationships between lipophilicity and root uptake and translocation of non-ionized chemicals by barley, *Pestic. Sci.*, 13: 495–504.

Burken, J.G., Schnoor, J.L., 1998, Predictive relationships for uptake of organic contaminants by hybrid poplar trees, *Environ. Sci. Technol.*, 32: 3379–3385.

Collins, C., Fryer, M., Grosso, A., 2006, Plant uptake of non-ionic organic chemicals, *Environ. Sci. Technol.*, 40: 45–52.

Costera, A., Feidt, C., Marchand, P., Le Bizec, B., Rychen, G., 2006, PCDD/F and PCB transfer to milk in goats exposed to a long-term intake of contaminated hay, *Chemosphere*, 64: 650–657.

DEFRA, 2002, Agency E "Potential contaminants for the assesment of land. Report CLR8," Available from The R&D Dissemination Centre, WRc plc.

Groom, C.A., Halasz, A., Paquet, L., Olivier, L., Dubois, C., Hawari, J., 2002, Accumulation of HMX (octahydro-1,3,5,7-tetranitri-1,3,5,7- tetrazocine) in indigenous and agricultural plants grown in HMX-contaminated anti-tank firing-range soil, *Environ. Sci. Technol.*, 36: 112–118.

Hsu, F.C., Marxmiller, R.L. Yang, A.Y.S., 1990, Study of root uptake and xylem translocation of cinmethylin and related-compounds in detopped soybean roots using a pressure chamber technique, *Plant Physiol.*, 93: 1573–1578.

Kierkegaard, A., Asplund, L., de Wit, C.A., McLachlan, M.S., Thomas, G.O., Sweetman, A.J., Jones, K.C., 2007, Fate of higher brominated PBDEs in lactating cows, *Environ. Sci. Technol.*, 41: 417–423.

Loutfy, N., Fuerhacker M., Tundo, P., Raccanelli, S., Ahmed, M.T., 2007, Monitoring of polychlorinated dibenzo-p-dioxins and dibenzofurans, dioxin-like PCBs and polycyclic aromatic hydrocarbons in food and feed samples from Ismailia city, Egypt, *Chemosphere*, 66: 1962–1970.

Mamontova, E.A., Tarasova, E.N., Mamontov, A.A., Kuzmin, M.I., McLachlan, M.S., Khomutova, M.I., 2007, The influence of soil contamination on the concentrations of PCBs in milk in Siberia, *Chemosphere*, 67: S71–S78.

McLachlan, M.S., 1993, Mass Balance of Polychlorinated-Biphenyls and Other Organochlorine Compounds in a Lactating Cow, *J. Agric. Food Chem.*, 41: 474–480.

McLachlan, M.S., 1996, Bioaccumulation of hydrophobic chemicals in agricultural feed chains, *Environ. Sci. Technol.*, 30: 252–259.

McLachlan, M.S., 1999, Framework for the interpretation of measurements of SOCs in plants, *Environ. Sci. Technol.*, 33: 1799–1804.

McLachlan, M., Richter, W., 1998, Uptake and transfer of PCDD/Fs by cattle fed naturally contaminated feedstuffs and feed contaminated as a result of sewage sludge application. 1. Lactating cows, *J. Agric. Food Chem.*, 46: 1166–1172.

Meneses, M., Schuhmacher, M., Domingo, J.L. 2002, A design of two simple models to predict PCDD/F concentrations in vegetation and soils, *Chemosphere*, 46: 1393–1402.

Ryan, J.A., Bell, R.M., Davidson, J.M., Oconnor, G.A., 1988, Plant uptake of non-ionic organic-chemicals from Soils, *Chemosphere*, 17: 2299–2323.

Seuntjens, P., Sterbaut, W., Vangronsveld, J., 2006, "Chain model for the impact analysis of contaminants in primary food products," Belgian Science Policy Secretariat, Rue De La Science, Brussels.

Shaw, S.D., Brenner, D., Berger, M.L., Carpenter, D.O., Hong, C.S., Kannan, K., 2006, PCBs, PCDD/Fs, and organochlorine pesticides in farmed Atlantic salmon from Maine, eastern Canada, and Norway, and wild salmon from Alaska, *Environ. Sci. Technol.*, 40: 5347–5354.

Simonich, S.L., Hites, R.A., 1994, Vegetation-atmosphere partitioning of polycyclic aromatic-hydrocarbons, *Environ. Sci. Technol.*, 28: 939–943.

Traag, W.A., Kan, C.A., van der Weg, G., Onstenk, C., Hoogenboom, L.A.P., 2006, Residues of dioxins (PCDD/Fs) and PCBs in eggs, fat and livers of laying hens following consumption of contaminated feed, *Chemosphere*, 65: 1518–1525.

Trapp, S., Matties, M., 1995, Generic one compartment model for the uptake of organic chemicals by foliar vegetation, *Environ. Sci. Technol.*, 29: 2333–2338.

Welsch-Pausch, K., McLachlan, M.S., 1998, Fate of airborne polychlorinated dibenzo-p-dioxins and dibenzofurans in an agricultural ecosystem, *Environ. Pollut.*, 102: 129–137.

Welsch-Pausch, K., McLachlan, M.S., Umlauf, G., 1995, Determination of the principal pathways of polychlorinated dibenzo-p-dioxins and dibenzofurans to lolium-multiflorum (Welsh Ray Grass), *Environ. Sci. Technol.*, 29: 1090–1098.

DAIRY LIVESTOCK EXPOSURE TO PERSISTENT ORGANIC POLLUTANTS AND THEIR TRANSFER TO MILK: A REVIEW

STEFAN JURJANZ, GUIDO RYCHEN, CYRIL FEIDT
UR Animal et Fonctionnalités des Produits Animaux, ENSAIA-INRA de Nancy, France

Abstract: The main emission sources of Persistent Organic Pollutants (POPs) should be attributed to Human activities although certain natural events may also enhance their production. After emission, these compounds can be potentially transferred to the food chain via interactions with livestock systems. POPs are characterized by some volatility, strong persistence in the environment, and a high lipophilicity, which leads to their accumulation in fat tissues. These molecules have raised concern about the risk of transfer through the food chain via the animal product. POPs are listed in several international conventions dealing with their potential toxicity for humans and the environment. This paper synthesizes current information on dairy ruminant exposure to POPs and the risk of their transfer to milk. Five major families of POPs have been considered: the two groups of dioxins and furans (PCDD/Fs), the Poly Chloro Biphenyls (PCBs), the Polycyclic Aromatic Hydrocarbons (PAHs) and the emerging family of Poly-Bromo-Diphenyl-Ethers (PBDE). Dairy ruminants are mainly exposed to theses POPs by oral ingestion, other contamination ways are considered as minor. The contamination of roughage and soil by these compounds is observed when they are exposed to emission sources (steelworks, cement works, waste incinerators or motorways) compared to remote areas. Concentrations in soil can be higher than in plants, especially for very persistent halogened compounds. Highest concentrations of POPs in soil may be close to 1 µg per kg dry matter for PCDD/Fs and PBDEs, 100 µg/kg dry matter for indicator PCBs, 10 g/kg dry matter for PAHs. The contamination of milk by persistent organic pollutants depends on environmental factors, which are related to the rearing system (feeding system, access to contaminated soil or not, stage of lactation, udder health of the animals) and of the characteristics of the considered contaminants (their chemical properties like molecular weight, halogenations, lipophilicity and metabolic susceptibility). Established transfer rates to milk were lowest for PAHs (generally less than 1%), mainly due to other excretion pathway like urine but also to a certain proportion of metabolization. Transfer rates of halogenized compounds were generally higher: for PCBs the rate of transfer varies widely from 5% to 90%, and for the PCDD/Fs from 1% to 40%. Further studies should clarify the effect of hydroxyl-metabolites of PAHs on Human health and precise the transfer rates of PBDEs where only few data are available.

Keywords: Pollutant transfer, contaminant in milk, soil contamination, fodder contamination

1. Introduction

Persistent Organic Pollutants (**POPs**) became in the last 30 years a topic of concern due to an increasing need of food safety all over the world. Indeed, several exceeding of regulatory thresholds in food, and especially in milk, in the last 10 years let emerge the need to understand better the way of introduction of such pollutants in the food chain.

POPs are mainly of an anthropogenic origin and their clearly negative effects on Human health make them be considered as pollutants. The persistence of POPs was defined by their half-life times. Indeed, the US Environmental Protection Agency (US EPA) defined the level of 120 days in a given matrix to consider a compound as persistent. Therefore an exposure to these compounds can be chronic, as they will remain a long time in the contaminated matrix.

This review synthesises scientific articles describing exposure and transfer of the main families of such contaminants into milk in order to improve the understanding the contamination pathways. This synthesis is thought to be a help step for monitoring livestock production systems in exposed situations.

B. Faye and Y. Sinyavskiy (eds.), *Impact of Pollution on Animal Products.*
© Springer Science + Business Media B.V. 2008

2. Chemical Properties of POPs (see figures in annexe)

Five major families of compounds should be considered today. The first family of contaminants is the ParaChloroDibenzo-para-Dioxins (**PCDDs**). These compounds are composed of two phenyl rings linked with two oxygen bridges and chlorinated from 4 to 8 times as shown in Figure 1a. The term PCDDs is grouping over all 75 compounds differing by the number and the position of chlorines but generally 7 main compounds were studied in the literature.

A second family of POPs is the ParaChloroDibenzoFurans (**PCDFs**). Two phenyl rings also composed them, but their link is composed by only one oxygen linking (Figure 1b). Chlorination of these composed ranked also from 4 to 8 and depending on their number and positions. Studies consider generally 10 main compounds among the 135 known congeners.

The first two families behave similarly and are commonly grouped in studies as **PCDD/Fs**, named also dioxins. These compounds are studied since the accident of Seveso (1976) and appear mainly during incomplete combustion processes with are generally due to human activities, especially all kinds of incinerators. Their half-life times are very high, especially in soil, and rank from 17 years (123678-HxCDD) to 274 years (OCDD) (Sinkkonen and Paasivirta, 2000). The most dangerous reference congener, 2,3,7,8 TCDD, was characterized by a half time of 41 years in soil.

PolyChloroBiphenyls (**PCBs**) are a third family of POPs composed of two phenyl rings linked by one direct linkage without any oxygen (Figure 1c). There are 209 recorded compounds totally (US EPA, 2003) but studies focused generally on different subgroups. The most often used group is the seven indicator congeners (PCB-I), which is composed by PCBs 28, 52, 101, 118, 138, 153 and 180. Another classification grouped four coplanar congeners (PCB 77, 81, 126 and 169) as well as eight non coplanar congeners (105, 114, 118, 123, 156, 157, 167, and 189) to the 12 dioxin-like PCBs (**PCB DL**). These compounds were manufactured as flame-retardants but their hazardous effect on human health lead to abandon their production in the seventies, especially in developed countries. The half-life times for PCBs in soil ranked from 3 (PCB 118) to 15 years (PCB 180), explaining that exposure to these POPs will continue to be relevant although the abandon of their synthesis. Their high persistence makes that we have to cop with PCBs for a long time in our environment.

Polycyclic Aromatic Hydrocarbons (**PAHs**) are the fourth family of POPs taken into account in our synthesis and are composed by 2–6 aromatic cycles (phenyl or furan rings) without any halogens (Figure 1d). The US EPA established a list of 16 priority compounds to follow but the European Union focused the priority only on 15 compounds exceeding three cycles (Directive 208/2005/EU). PAHs appear also during incomplete combustion what can happen during natural events as forest fires or volcanic eruptions. Nevertheless, the main origin of these compounds is anthropogenic activities as incinerators or domestic heating, highly frequented traffic ways (motorways, airports) or industrial activities (coke manufacture, chemical and petrochemical industries). PAHs can reach half-life times in soil over 100 days when the number of rings in the molecule exceeds four and can therefore easily reach the persistence level indicated by the US EPA.

Finally, an emerging family of compounds – PolyBromoDiphenylEthers (**PBDEs**) – are composed by two phenyl rings linked with one oxygen bridge, more or less intensively brominated with 1 (MonoBDEs) to 10 (DecaBDEs) ions (Figure 1e). In theory, there could be as many as 209 congeners. Generally, only 11 congeners brominated 3–10 times and three types of commercial products (Penta-, Octa- and DecaBDEs) are generally of concern. This more recently considered family is always produced voluntary as a flame-retardant but their production run also down in developed countries. Their half-life times are generally lower. Indeed, highly brominated forms (Deca- and Nona BDEs) can be partially debrominated within 2–3 weeks. Nevertheless, PBDEs can also resist to degradation over long periods when they are protected against degradation within an environmental matrix. Indeed, revealed concentrations of the most dangerous Penta- and HexaBDEs show the elevated persistence in such matrixes.

Thus, the low degradation rate of POPs will maintain these contaminants a long time in the environment and will therefore cause mainly chronic exposure to animals and humans. All POPs are characterized by a more or less elevated lipophilicity. This biochemical property was generally expressed as the partition coefficient octanol-water (**Kow**, generally used in the Log form) corresponding to the ratio between the possible concentration of a considered POP in octanol and its possible concentration in water. Indeed, the Log Kow of PAHs increases with the number of rings in the compound from 3,4 (Naphtalene) to 7,2 (Benzo(g,h)perylen). The lipophilicity of PCDD/Fs is firstly higher than in PAHs and secondly increases with the number of chlorines in the congener between 6,8 (2378 TCDD) and 8,8 (OCDF). This lipophilicity explains their ability to be stored and accumulated in fat.

3. Exposure of Dairy Livestock to POPs

The sources of pollution correspond to three groups according to Wild and Jones (1995) and Lichfouse et al. (1994, 1997): industrial activities (energy production, metallurgy, cement works, chemical industries,…), urban activities (transport, management and processing of waste) and husbandries (mud spreading, domestic heating). In addition to these stationary sources, PAHs are also emitted by mobile sources as highly frequented motorways or airports.

The industrial activities that are most polluting in PAHs and chlorinated compounds are those using fossil fuels (Edwards, 1983; Kakareka, 2002; Krauss and Wilcke, 2003). Another source of contamination of the soil is the production of coke from coal. Indeed, the mechanisms of condensation, decantation and distillation of the tar from the furnace (tar being a by-product of the manufacture of coke) generate the formation of organic pollutants, which are then emitted into the atmosphere. In addition, soil contamination by PAHs, PCDD/Fs and PCBs is often concomitant with those of metal pollutants. Stalikas et al. (1997) showed that lignite combustion generates large quantities of PAHs. In the same way, PCDD/Fs and PCBs are released in the atmosphere bordering the sites of cement works. Henner (2000) also points out that another source of soil pollution by PAHs is the sites of gas extraction.

Indirect effects such as the degeneration of the plants may increase these direct contaminations of soil by atmospheric emissions, of anthropogenic origin. Indeed, $44 \pm 18\%$ of the atmospheric PAHs are introduced into the soil following their capture on waxy surfaces of plants followed by the decomposition of the plant (Simonich and Hites, 1994). Finally, the pollution of soil can also result from the natural contribution of organic pollutants (for example, the pyrolysis of humus during forest fires) (Laurent et al., 2005).

POPs are transported from the emitting sources mainly by air. The form of transport (in the gas fraction or adsorbed at particulates) and the distance before deposition is depending on the chemical properties of the considered compound. After transport, the POPs are deposited in the environment.

Gas deposit concerns the most volatile compounds, namely the least chlorinated PCBs and the low molecular weight PAHs with two or three aromatic cycles (Howsam et al., 2000). The least volatile compounds are found mainly in the form of particulate deposit (Welsch-Pausch et al., 1995). Thus, PCDD/Fs are found deposited in particulate form, PCBs are reported to be found primarily in gas form (Thomas et al., 1998) and the PAHs are found in one form or the under, or in the two forms according to their partition coefficient octanol-air (**Koa**).

Contrarily, PBDEs seem to enter in the food chain by the aquatic way via waste-waters of manufactures or discharges of electronic wastes. Nevertheless PBDEs are strongly attracted by the organic matter as sediments, sewage sludge or soil allowing them to enter in the terrestrial food chain and to get in contact with livestock systems as shown by Fangström et al. (2005).

The different ways of contamination of the dairy ruminant by POP are ingestion of polluted matrixes, inhalation of contaminated air or absorption by dermal contact. In the lactating animal, the exposure by inhalation is considered as negligible when compared to oral administration of contaminated feed or soil. The skin absorption of organic pollutants was not studied, but several studies undertaken on laboratory animals suggest that this exposure is also negligible under the conventional conditions of breeding. Therefore, research work has been focused towards the contamination of the ruminant via the feed way, i.e. the ingestion of feed or soil. Industrial processed concentrates can be contaminated. Moreover, vegetation and soil can be exposed to the deposit of POPs before getting in contact with animals of livestock systems, especially during grazing.

3.1. EXPOSURE TO POPs VIA SOIL

A lactating ruminant may ingest daily from 1% to 10% soil when grazing (Healy, 1968; Thornton and Abrahams, 1983). This data were measured mainly in cattle and nearly no information is available concerning the intake in another dairy species. Thus, the soil intake during grazing should depend on the climatic conditions on pasture and ingestion behaviour of the considered specie. Indeed, the pasture in arid conditions let increase the daily intake of soil over 1 kg (Mayland et al., 1975; Kirby and Stuth, 1980). Cattle pull up tufts when graze what make them especially exposed to soil intake. Sheep and goats have a similar behaviour of grass intake what let suppose the same exposure in these species. This question is quite different in camels as they thin out the leaves of branches (Jarrige et al., 1995) and mares would cut the grass with his teeth what should considerably reduce the amount of soil ingested at pasture. Nevertheless, the veterinary experience show that considerable amounts of soil were occasionally found in stomachs during chirurgical interventions what make believe in a possible intake even in these species. Nevertheless, no data is available to quantify this intake and to estimate possible exposure thereafter.

Soil contamination by organic pollutants occurs primarily by atmospheric deposit (Laurent et al., 2005) and many authors have highlighted the fact that the concentration of POPs in soil increases with the density of the human activities. When deposited on the surface of the soil, these compounds tend to remain in the superficial surface (the first 15 cm of soil; Fries, 1982; Steven and Gerbec, 1988; Jones et al., 1989).

Concentrations of POPs in different soils are given in Tables 1–3. Firstly, highest concentrations of PCDD/Fs in soil reported in the literature may be close to 1 µg/kg dry soil (Table 1), although they rarely exceed 500 ng/kg. Octa chlorinated congeners are by far the most represented compound, but the very small proportion of tetra chlorinated forms should not hide the fact that TCDD/F are the most toxic forms. Only few studies reported concentrations of PBDEs in soil (Hassanin et al., 2004; Law et al., 2006). They seem to reach values close to these revealed for PCDD/Fs, i.e. up to 1 µg/kg. Data on concentration of PCBs in soil are very variable and even focusing of indicator PCBs, the reported concentrations varied widely (Table 2). It seems to us that concentration of 100 µg of indicator PCBs/kg dry soil would be exceeded and only in soils not really relevant for agricultural purposes. Finally, the concentrations of PAHs in very exposed sites can reach up to 10 mg/kg dry soil (Table 3). Although the profiles vary depending on emission source and characteristics of the site, the major compounds are generally phenanthrene, fluoranthene and pyrene. Nevertheless a special attention should always be paid to benzo[a]pyrene as it is the most dangerous congener.

TABLE 1. Reported contamination of soils in PCDD/Fs (ng/kg dry soil)

	Rural area	Cement plant	Near waste incinerator		
2,3,7,8 TCDD	0.3	0.03	0.11	0.41	0.09
1,2,3,7,8 PeCDD	1	0.09	0.45	1.31	0.4
1,2,3,4,7,8 HxCDD	1	0.12	0.52	1.48	0.68
1,2,3,6,7,8 HxCDD	3	0.2	1.25	4.03	1.61
1,2,3,7,8,9 HxCDD	4	0.2	0.8	4.23	1.86
1,2,3,4,6,7,8 HpCDD	22	3.7	15.4	46.9	31.5
OCDD	89	25.3	75.0	761	139
2,3,7,8 TCDF	9	0.25	1.32	12.74	0.93
1,2,3,7,8 PeCDF	4	0.15	0.45	2.23	1.03
2,3,4,7,8 PeCDF	4	0.17	1.27	4.73	1.07
1,2,3,4,7,8 HxCDF	6	0.18	1.82	10.6	1.61
1,2,3,6,7,8 HxCDF	5	0.2	1.63	3.27	1.54
1,2,3,7,8,9 HxCDF	4	0.04	0.53	0.23	3.07
2,3,4,6,7,8 HxCDF	0.4	0.22	2.02	4.47	0.14
1,2,3,4,6,7,8 HpCDF	23	1.05	8.03	21.2	13.7
1,2,3,4,7,8,9 HpCDF	3	0.08	1.33	2.44	1.66
OCDF	30	0.73	11.8	21.5	16.75
Total PCDD/Fs	**209**	**33**	**124**	**903**	**217**
Reference	(1)	(2)	(3)	(4)	(5)

(1) Hassanin et al. (2005); (2) Schuhmacher et al. (2002); (3) Blanchard et al. (2001); (4) Domingo et al. (2002); (5) Floret et al. (2007)

TABLE 2. Reported concentrations of indicator PCBs in soil (µg/kg of dry soil)

Type of site	Revealed concentration	Reference	Remarks of the authors
Not polluted	<0.7	Schumacher et al. (2004)	
Petrochemical site	2–5	Manz et al. (2001), Schumacher et al. (2004), Nadal et al. (2006)	
Urban area	10	Nadal et al. (2006)	
Chemical site	20	Schumacher et al. (2004)	
Siberian pasture	92	Iwata (1995)	Amazingly elevated
Waste discharge	180	Minh et al. (2006)	
Agricultural soil	215	Alcock et al. (1997)	Amazingly elevated
Industrial site	500	Turrio et al. (2007)	Probably very close to the emitting source

TABLE 3. Reported concentrations of PAHs in soil (mg/kg of dry soil)

	Near highway	Urban area	Forest soil	Near coke manufacture	Near gas manuf.	Netrochimic industry
Naphtalene	15	6	154	na	na	1,131
Acenaphtene	4	37	86	227	2	na
Acenaphtylene	12	na	na	na	na	333
Fluorene	5	na	na	na	225	650
Phenanthrene	76	14	533	379	379	1,595
Anthracene	13	2	117	156	1156	334
Fluoranthene	162	27	1,132	2,174	2,174	682
Pyrene	123	31	573	4911	91	642
Benzo[a] anthracene	79	20	796	662	317	na
Chrysene	126	–	–	–	345	614
Benzo[b] fluoranthene	163	10	492	92	2271	na
Benzo[k] fluoranthene	54	na	na	na	–	na
Benzo[a]pyrene	70	7	352	260	92	na
Indenol[1,2,3-cd] pyrene	3	na	na	na	120	na
Benzo[g,h,i] perylene	102	33	685	na	na	na
Dibenzo[a,h] anthracene	48	na	na	na	192	na
Total PAHs	**1,145**	**187**	**4,920**	**8,861**	**7,364**	**5,981**
Reference	Crépineau et al. (2003)	Wild and Jones (1995)			Juhasz and Naidu (2000)	

na : non analyzed

Because of their lipophilic character and a low aqueous solubility, there is a strong POP adsorption in the organic matter of the soil. The rinsing of the soil POPs is thus regarded as negligible (US EPA, 2000).

The degradation of POPs in soil can be carried out by photodegradation and microbial degradation (Laurent et al., 2005). These phenomena generate the appearance of metabolites, with a change in the chemical structure. This causes modifications of their toxicity (in particular for the PAH) and of their behaviour in the soil compared to that of the parent compounds (Schiavon, 1988). The photodegradation or abiotic degradation of the organic pollutants can take place only for the compounds located at the soil surface (Hebert and Miller, 1990), and especially the de-brominating of highly brominated PBDEs has been shown (Hassanin et al., 2004). PAHs can be degraded by photooxydation and reactions of oxidation (Juhasz and Naidu, 2000). They react with ozone to form quinones and epoxides. This mechanism is relatively important and allows a clear reduction in soil PAHs. The mechanism of photodegradation of the

PCDD/Fs and the PCBs "dioxins-like" has been the subject of some studies (Moore and Ramworthy, 1984; Dougherty et al., 1993; McPeters and Overcash, 1993). It implies a dechlorination (Helling et al., 1973), the slightly chlorinated molecules being more easily photolysed than the octa-chlorinated compounds (Helling et al., 1973; Dougherty et al., 1993). However, this mechanism would not be very important since soil limits the penetration of ultraviolet radiation. However Miller et al. (1989), Kieaitwong et al. (1990) and Tysklind et al. (1992) highlighted that the chlorine atoms, in a *peri* position, from the strongly chlorinated PCDD were eliminated, leading thus generally to the formation of the 2,3,7,8-TCDD.

The microbial or biotic degradation of the organic pollutants of the soil is a mechanism controlled by the temperature, by the properties of the soil (water and or-ganic matter content, pH) and the compounds (molecular weight and Log Kow) (Cerneglia, 1992; Bakker and de Vries, 1996; Mhiri and de Marsac, 1997). These activi-ties of POP degradation in the soil play a role in the carbon cycle. PAHs are composed of carbon and hydrogen and thus form an integral part of this cycle (Gibson and Subramanian, 1984). Indeed, their similarity with other organic molecules means that telluric micro-organisms have the enzymatic ability to degrade them. Thus, microbial degradation is a more important process of decontamination of soils than the process based on the photo-oxidation or self-oxidation. In a general way, the mechanism is accelerated in the presence of nutrients added to the soil (Wilson and Jones, 1993; Straube et al., 1999), of organic matter (Kästner and Mahro, 1996), of a ventilation of the soil and an increase in temperature (Bonten et al., 1999), these different factors probably supporting the development of micro-organisms.

Microbial degradation of PCBs is carried out according in one of two ways: the strongly chlorinated compounds can undergo a dechlorination in anaerobic conditions while the others generally undergo an oxidation by bacteria developing in aerobiosis (Abramowicz, 1990). The importance of this microbial degradation is still being discussed. According to Mhiri and de Marsac (1997) the biological breakdown of soil PCBs allows a significant elimination of these molecules, whereas for Sierra et al. (2003), these same molecules are slightly metabolized. It is the same for the PCDD/F: in a general way, the degradation of the PCDD/F by bacteria of the soil is regarded as a rather ineffective way of dissipation, requiring many years and particularly when the compounds are strongly chlorinated (Arthur and Frea, 1989; Paustenbach et al., 1992; Beurskens et al., 1995; Wittich, 1998). Habe et al. (2001, 2002) highlighted that, after 7 days of inoculation with *Terrabacter* sp. (stock DBF63) the rate of degradation of the molecules with 4–6 chlorine atoms was close to 10%, whereas for the highest chlori-nated compounds it was close to zero.

3.2. EXPOSURE TO POPS VIA PLANTS

The plant being the interface between the soil and the atmosphere, its contamination is likely to occur either by atmospheric deposit, or by root absorption. Four ways of entry of POPs have been distinguished (Welsch-Pausch et al., 1995; Bakker et al., 2001; Teil et al., 2004): gas deposit, dry deposit of particles, wet deposit of particles and root absorption. Contaminations by root absorption are considered by many authors as negligible (Wild and Jones, 1992; Welsch-Paush et al., 1995; Kipopoulou et al., 1999) since the POPs are not very soluble but are very lipophilic compounds (Simonich and Hites, 1994). Each mode of contamination must be taken into account in order to evaluate the entry of the pollutants into the plant, their availability for the ruminant, and to characterize and model the contamination of fodder. Generally, the concentration in the plants depends on the distribution and nature of the compounds in the atmosphere and their properties influencing the proportion between particulate and gas form.

Environmental conditions, characteristics of the plants and physicochemical properties of the compounds represent numerous factors, which may influence the type of deposit and the quantity of pollutants found on the plant. The main environmental conditions influencing the content of POPs in plants are temperature, rainfalls and wind. Temperature affects the form in which the POP compounds are present in the atmosphere: gas or particles (Howsam et al., 2000; Bakker et al., 2001; Blais et al., 2003). Some compounds are thus present in the atmosphere in particulate or gas form according to the ambient temperature (Koa of the compounds is temperature-dependent): this is the case for example of the TCDD, which is exclusively in gas form during the summer (Welsch-Pausch and McLachlan, 1998). The concentration of the pollutants in vegetation also depends on temperature: it increases by a factor of 30–2,000 when the temperature increases from 5°C to 50°C (Bakker et al., 2001). Wind speed and direction also affect the concentration of POPs in plants (Bakker et al., 2001; Lohman and Seigneur, 2001; Teil et al., 2004; Smith et al., 2001) by modifying the distribution of the compounds in the atmosphere. The role of rainfalls on the rinsing of the compounds depends on their nature and the plant concerned: for example, a lettuce washed with water involves a significant extraction of high molecular weight PAHs, however water will extract only a weak part of high molecular weight of PAH or PCDD/F in corn (Bakker et al., 2001).

POP concentrations in plants are also dependent on the characteristics of the plant such as the pilosity of the leaf, the composition of the cuticle or the architecture of the plant. The cuticle, rich in waxes can involve an increase in the accumulation of the lipophilic molecules (Müller et al., 2001). Cutin waxes composing the cuticle is responsible for 70–90% of adsorption (Thomas et al., 1998) and the quality of their waxes present rather than the thickness of the cuticle, which modifies the concentration (Smith et al., 2001). There is a passive diffusion between the atmosphere and the cuticle in the case of gas deposits. The molecules are transferred until an air-plant balance is reached. The time to reach balance varies according species: from 24 to 240 seconds for the *Citrus* species and from 58 to 580 days for the species of *Ilex* (Bakker et al., 2001). Thus for certain plants, a balance is never reached because the lifespan of the plant is too short (Thomas et al., 1998; Smith et al., 2001). Furthermore, the surface of deposit of the plants is also a major factor affecting POP deposit: it is for example 6–14 times higher than that of the soil on which they develop (Simonich and Hites, 1994).

The physicochemical characteristics of the various POPs are also among the principal factors influencing the contamination of grass. The levels of PCDD/Fs measured in grass rarely exceed 50 ng/kg dry grass and these for indicator PCBs may attain around d 1 µg/kg in exposed sites (Table 4).

No data are available concerning levels of PBDEs in grass or similar terrestrial vegetation. Finally, the literature reported much higher levels of PAHs in grass. Indeed,

TABLE 4. PCDD/F and PCB levels in grass samples collected near a cimentory or a hazardous waste incinerator (ng/kg dry matter)

| | Type of site | | | | Type of site | | |
Compounds	Near a cement plant	Industrial area	Near a hazardous WI	Compounds	Control grass (remote area)	Near a hazardous WI
2,3,7,8 TCDD	nd	2.8	0.06	PCB 77	3.86	10.43
1,2,3,7,8 PeCDD	0.04	<0.1	0.65	PCB 81	0.14	0.59
1,2,3,4,7,8 HxCDD	0.08	0.73	0.64	PCB 126	0.63	3.02
1,2,3,6,7,8 HxCDD	0.10	6.0	0.68	PCB 169	0.12	1.03
1,2,3,7,8,9 HxCDD	0.06	4.2	0.61	PCB 105	19.45	42.42
1,2,3,4,6,7,8 HpCDD	0.62	13	4.65	PCB 114	1.27	2.49
OCDD	1.68	43	10.33	PCB 118	63.69	126.7
2,3,7,8 TCDF	0.21	16	0.35	PCB 123	2.23	5
1,2,3,7,8 PeCDF	0.11	1.2	0.62	PCB 156	6.68	13.82
2,3,4,7,8 PeCDF	0.11	<0.09	1.00	PCB 157	1.17	2.87
1,2,3,4,7,8 HxCDF	0.11	4.6	1.18	PCB 167	4.57	8.26
1,2,3,6,7,8 HxCDF	0.10	1.8	1.37	PCB 189	0.9	2.22
1,2,3,7,8,9 HxCDF	nd	0.54	0.20	PCB 28	57	66.76
2,3,4,6,7,8 HxCDF	0.10	2.4	1.73	PCB 52	46	68.53
1,2,3,4,6,7,8 HpCDF	0.39	11	4.69	PCB 101	169	215.57
1,2,3,4,7,8,9 HpCDF	0.06	0.89	0.35	PCB 138	176	283.42
OCDF	0.43	8.0	1.49	PCB 153	319	457.98
				PCB180	68	126,45
Sum PCDD/Fs	**4.2**	**116**	**31**			
Reference	(1)	(2)	(3)		(3)	

nd: non detected
(1) Schumacher et al. (2002); (2) Jones and Davidson (1997); (3) Costera et al. (2006)

TABLE 5. PAH levels in grass samples (mg/kg dry grass)

Plants	Concentration	Compounds	Sampling sites	Reference
Grass	142	Σ16 PAH	Road (13,000 cars/day)	Müller et al. (2001)
Grass	1,461	Σ16 PAH	Highway (F)	Bryselbout et al. (2000)
Ray-grass	100–900	Σ24 PAH	Rural area (UK)	Smith et al. (2001)
Fescue	136–510		Rural area (UK)	
Grass	153	Σ16 PAH	Urban area (UK)	Meharg et al. (1998)
Grass	900	Σ16 PAH	Highway (F)	Crépineau et al. (2003)
Grass	25	Σ16 PAH	Remote pasture (F)	Crépineau-Ducoulombier and Rychen (2003)
Plantain	200–1,700	Σ8 PAH	Urban area (NL)	Bakker et al. (2000)

concentration of up to 1 g/kg dry matter were reported (Table 5) concerning highly exposed sites. The proportion of the gas or particulate deposit depends on the characteristics of the molecules (Howsam et al., 2000) and in particular of their volatility (Bakker et al., 2001), their lipophilicity measured by the coefficient Koa, their solubility in water, their steam pressure value, their constant of Henry (Meneses et al., 2002) and their half-life (Kipopoulou et al., 1999). The value of Koa varies according to the compounds and directly influences gas-particles distribution (Lohman and Jones, 1998). When the Koa is high, the particulate deposit increases.

For example, the PCBs with high chlorination have a low steam pressure and high Koa and thus are found in particulate form (Jan et al., 1994). The PCDD/Fs having six chlorine atoms or more are also in particulate form (Bakker et al., 2001). The PAHs with two and three cycles are exclusively in gas form, those of more than five cycles are exclusively in particulate form. The PAH with four cycles are distributed between the two forms according to the ambient temperature (Howsam et al., 2000). This great variability is important for a better understanding of the profiles detected in the plants. For example, 95% of phenanthrene is always in gas form whereas this value is less than 10% for benzo(ghi)perylene (Wild and Jones, 1992).

The amount of ingested fodder, possibly contaminated, depends on the considered species in the livestock system (Table 6). The daily intake on pasture would be the most relevant in our context and it can reach 80 kg of fresh grass corresponding to 15 kg dry matter (**DM**).

A final comparison of reported concentrations of POPs in both environmental matrixes (Table 7) shows that both matrixes will contribute to the POPs exposure. Highly persistent PCDD/Fs and PCBs can be 100-fold more accumulated in soil during several years (memory effect of soil) what makes them a more relevant vector of such POPs to grazing livestock systems than plants accumulating such POPs only during one vegetation period. Moreover, soil can protect PCDD/Fs more efficiently against abiotic degradation. The exposure of dairy ruminants to parent PAHs concerns soil and fodder in a similar way. Although grass is generally less contaminated (approximately tenfold) than soil, the higher daily intake during grazing would make that animals will ingest comparable amounts of such compounds via soil and grass on exposed sites. Nevertheless, accidental access to other sources of exposure – generally less chronic but more punctual – can not be completely neglect.

TABLE 6. Daily intake of fodder and soil depending on species

Species	Fodder intake on pasture (kg DM/day)	Soil intake on pasture (kg/day)
Dairy cattle	15	Up to 1
Beef cattle	8	Up to 1
Sheep	1,5	
Goat	1,5	
Camel	6,5–10[a]	No data available
Mare	3/100 kg BW	Probably up to 10%[b]

[a] personnel communication, B. Faye

[b] simulation IAEA, G. Voigt

TABLE 7. Comparison of upper levels of contamination with POPs between roughage and soil when they are exposed

	Upper concentration reached in exposed soil	Upper concentration reached in exposed grass
PCDD/Fs	1 μg/kg	10 μg/kg dry matter
PBDEs	1 μg/kg	No data available
Indicator PCBs	100 μg/kg	1 μg/kg dry matter
PAHs	10 g/kg	1 g/kg dry matter

4. Transfer Rates of POPs to Milk

When considering the transfer of POP to milk, the PAHs and the halogened compounds need to be distinguished: PCDD/Fs and PCBs are generally considered as persistent and bioaccumulable in the livestock products whereas PAHs are considered as largely metabolized. Only few is known about transfer rates of PBDEs but they seem to behave similarly as chlorinated POPs. Although the physicochemical characteristics of PAHs are well described, their interaction with the metabolism of the dairy ruminant is not yet well known.

4.1. TRANSFER OF HALOGENED POPS TO MILK

Table 8 indicates the values of the PCDD/Fs transfer rates "feed-milk" determined by several authors.

The obtained values oscillate between 1% and 50% and all the individual PCDD/F compounds were found in milk. For the compounds whose log Kow is higher than 6.5, the transfer appeared to be reduced with an increase of this partition coefficient. Some exceptions however have been detected: among the PCDFs, the weak transfer of 2,3,7,8TCDF, 1,2,3,7,8PeCDF and 1,2,3,7,8,9 HxCDF may be related to the hepatic degradation of the compounds. The contents of PCBs in cow's milk were less studied: concentrations were usually in the range of 1 pg/g fat, except for the PCB 118 whose concentrations reached levels 1,000 times higher (Willett et al., 1987, 1989; McLachlan, 1993; Sewart and Jones, 1996; Krokos et al., 1996; Focant et al., 2003). In a recent study (Mamontova et al., 2007), the relationship between PCB levels in cow's milk and in pasture soil was assessed and demonstrated.

TABLE 8. Transfer rates of PCDD/Fs "feed-milk" reported in the literature

Compound	Cow	Cow	Cow	Goat	General evaluation
2,3,7,8 TCDD	35	15	35	53	Very high
1,2,3,7,8 PeCDD	33	10	28	33	Very high
1,2,3,4,7,8 HxCDD	17	5.6	18	24	Intermediate
1,2,3,6,7,8 HxCDD	14	6.4	16	25	Intermediate
1,2,3,7,8,9 HxCDD	18	3.1	12	15	
1,2,3,4,6,7,8 HpCDD	3	0.6	1.8	5.4	Low
OCDD	4	0.1	0.3	1.7	Very low
2,3,7,8 TCDF	nd*	nd	nd	10.2	
1,2,3,7,8 PeCDF	nd	nd	nd	14	
2,3,4,7,8 PeCDF	25	12	18	29	High
1,2,3,4,7,8 HxCDF	nd	4.3	5.7	22	
1,2,3,6,7,8 HxCDF	16	3.6	11	18	Intermediate
1,2,3,7,8,9 HxCDF	nd	nd	nd	3.0	
2,3,4,6,7,8 HxCDF	14	4.2	8.4	12.5	Intermediate
1,2,3,4,6,7,8 HpCDF	3	0.4	1.4	2.7	Low
1,2,3,4,7,8,9 HpCDF	8	0.5	nd	3.5	
OCDF	1	nd	0.1	0.9	Very low
Reference	McLachlan et al. (1990)	Slob et al. (1995)	Fries et al. (1999)	Costera et al. (2006)	

nd: non determined

In another recent study (Costera et al., 2006), the feed to milk transfer of 17 PCDD/Fs and 18 PCBs was established in lactating goats exposed during 10 weeks to the intake of contaminated hay collected in the vicinity of a hazardous municipal waste incinerator. For PCDD/Fs (Table 8), 2,3,7,8 TCDD appeared as the compound having the highest carry over rate (35%). For no coplanar dioxin-like PCBs, carry over rates higher than 80% were obtained for PCB 105, 118 and 157. Concerning indicator PCBs, the carry over rates ranged from 5% (PCB 101) to more than 40% (PCB 118, 153 and 180) (Figure 2). The intensity of the transfers appeared as a function of both the physico-chemical properties (especially chlorination) of the compounds and their metabolic behaviour according to Costera et al. (2006). This observation indicates a marked bio-transformation of the different compounds by the ruminant (Firestone et al., 1979; Rappe et al., 1987; McLachlan et al., 1990; Olling et al., 1991; Fries et al., 1999). According to Willett et al. (1989), the PCBs may be partially degraded during the fermentation of the ration in the rumen.

In milk produced in farms far from POP emission sources, the contents of PCDD/Fs were found between 1.3 and 2.5 pg I-TEQ/g of fat content (Rappe et al., 1987; Schmid and Schlatter, 1992; Eitzer, 1995; Harrison et al., 1996; Hippelein et al., 1996; Ramos et al., 1997). In exposed situations, the concentrations in PCDD/Fs were also generally lower than the threshold of 3 pg I-TEQ/g of fat content fixed in the European regulation (Directive 2375/2001). However, in some extreme situations, values of 9 pg I-TEQ/g of fat content have been reported (Costera et al., 2006), and in some situations of crises concentrations of 25 pg I-TEQ/g fat were reached for PCBs (confidential information).

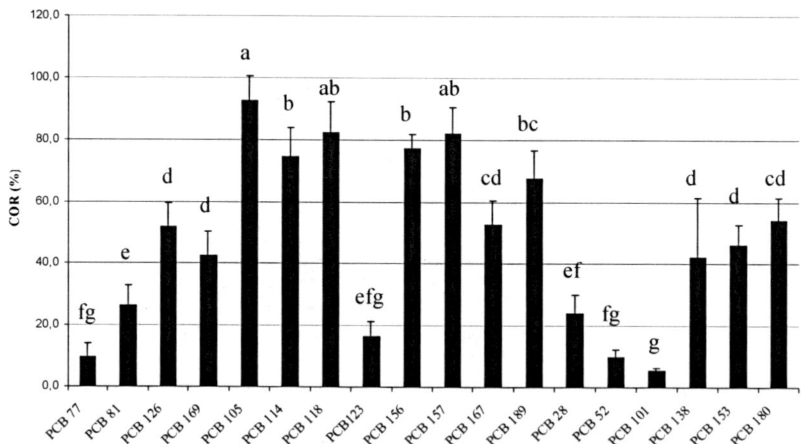

Figure 2. Transfer rates of PCBs from feed to milk according to Costera et al. (2006). a, b…g: different letters in a column indicates a difference significant at the threshold of $P < 0.05$

An increase in the contents of 2,3,4,7,8-PeCDF, 1,2,3,4,7,8-HxCDD/F, 1,2,3,6,7,8-HxCDD/F, 2,3,4,6,7,8-HxCDF, and 2,3,7,8-TCDD is noted around chemical, metallurgical industries or waste incinerators. The concentrations of dioxin-compounds in milk appeared to be linked to the number of chlorine atoms carried by the molecule (Ramos et al., 1997). The PCDD/F contents in milk are also dependent on rearing factors. Indeed, POP transfer to milk may fluctuate according to the physiological status of the animals. Tuinstra et al. (1992) showed for example that the disappearance of the PCDD/F in milk was linked to the body fat reserves of the animal. The mobilization of these reserves during the cycle of lactation can thus have a significant impact on the content of these compounds in milk. Indeed, at the beginning of lactation, animals do not manage to meet their needs via feed ingestion and therefore use their body fat reserves. Later during lactation, the opposite phenomenon (nutritive contribution in excess compared to the needs) occurs and POPs may be stored in the body fat reserves (Jarrige, 1988; Thomas et al., 1999). Thus, the peak of concentration of PCDD/Fs observed in the colostrum (Tuinstra et al., 1992) could be related to the contamination of the animal during its phase of reconstitution of the body reserves (during the previous lactation or the dry off period). Finally, Fries et al. (1999) noted an increase in the concentrations of PCDD/Fs strongly chlorinated during the infection of the mammary gland. This phenomenon can be explained by the structural modifications of the cells of the mammary gland during mastitis: increased permeability of the mammary epithelial barrier (Fries et al., 1999).

Milk levels of POPs can also fluctuate during the lactation according to seasons and according to the diets of the animals (Krokos et al., 1996). Indeed during the summer period when animals are on pasture), the involuntary ingestion of soil whose POP contamination may be elevated can represent a major concern (Fries and Paustenbauch 1990; Mamontova et al., 2007).

Only few studies focused on the PBDE transfer into milk. Different market surveys revealed concentrations of 162 pg/g of milk fat in Finland (Kiviranta et al., 2004) but concentrations around 500 pg/g of milk fat in two recent studies about dairy products in Spain (Bocio et al., 2003; Domingo, 2006). Moreover, epidemical studies on women confirmed above cited concentrations for Human milk in the USA (Betts, 2001) but others (Zurbier et al., 2006) reported strongly higher levels in milk of women in the Spain, Netherlands, UK and Germany of 2–8 ng/g of milk fat. These very high values should incite to study more precisely the transfer of PBDEs in the food chain although their TEFs are not determined yet.

4.2. TRANSFER OF PAHS TO MILK

Studies on transfer of parent PAHs in the terrestrial food chain appear limited owing to the fact that these compounds can be strongly metabolized. These molecules have nevertheless a recognized toxicity and their transfer to milk has been assessed in the recent years (Rychen et al., 2005). The aim of the studies was to characterize the transfer of the parent compounds and their main hydroxylated metabolites. First, milk samples coming from farms located near to emitting sources and "Control farms" distant from all potential sources of contamination (30 km) were collected for PAH analysis. The analyses revealed the presence of PAH in all collected milks, concentrations varying from 20 to 30 ng/g fat. On the 16 sought PAHs, only the following 10 were detected with less than four aromatic cycles were detected: naphthalene, acenaphtylene, acenaphtene, fluorene, phenantrene, anthracene, fluoranthene, pyrene, benzo[a]anthracene, and chrysene (Grova et al., 2002a). The presence of PAH in conventional milks led these authors to study the transfer towards the milk of some model compounds in controlled situation (Grova et al., 2002b). Thus, three radioactive ^{14}C-PAHs (phenanthrene, pyrene and benzo[a] pyrene) and a dioxin compound (2,3,7,8 TCDD) were orally administered to lactating goats. Results of this study indicated a significantly different behaviour between the 4 studied compounds. Based on the share of radioactivity initially introduced and excreted in milk, three levels of excretion were observed: the TCDD with a rate of 7.8%, phenanthrene and pyrene with respectively 1.5% and 1.9% and benzo[a]pyrene with 0.2%. The rate of radioactivity associated with benzo[a]pyrene, phenanthrene and pyrene in urines (respectively 6%, 11%, and 40%) suggests a consequent biotransformation. This study showed for the first time that the radioactivity related to PAHs distributed by oral way to ruminants in lactation is actually transferred towards milk.

More recent work (Grova et al., 2006) made it possible to specify the phenomenon of biotransformation of PAHs within the organism and the distribution of the native molecules and their metabolites in excretion products. Lapole et al. (2007) demonstrated that recovery rates of pyrene and phenanthrene in milk appeared to be very low whereas the transfer of their corresponding metabolites was significantly higher (Table 9). Recovery rates in urine were found to be significantly higher (1–10 times) than recovery rates in milk. The 1-OH-pyrene was found to be the main metabolite in urine as well as in milk. Thus, this metabolite can be considered as a marker of exposure to PAHs also in ruminants. Furthermore, PAH metabolites should be taken into consideration when evaluating the safety of milk and further research should be conducted on these compounds. Lapole et al. (2007) also noticed that benzo[a]pyrene and 3-OH-benzo[a]pyrene were transferred to milk and urine in very small amounts (less than 0.005%). This very limited transfer rate suggests a low risk of exposure by humans to benzo[a]pyrene or its major metabolite from milk or dairy products.

TABLE 9. Recovered proportion of ingested PAHs transferred to milk and urine during 24 h (Lapole et al., 2007)

	Milk (%)	Urine (%)
Phenanthrene	0.014a	0.03a
Pyrene	0.006a	0.02a
Benzo[a]pyrene	0.002a	0.005a
3-OH-phenanthrene	0.073a	0.24a
1-OH-Pyrene	0.44b	1.175b
3-OH-Benzo[a]pyrene	0.001a	0

a, b: different letters in a column indicates a difference significant at the threshold of P < 0.05

One of the research perspectives remaining is to better characterize the bio-availability of PAHs from feed matrices or soil. Indeed, in most of the experiments (Grova et al., 2005; Lapole et al., 2007) PAHs were administered orally to lactating animals in oil samples. However, Lutz et al. (2006) determined the transfer kinetics of soil bound PAHs (fluorene, phenanthrene, pyrene and benzo[a]pyrene) to milk in lactating cows during a chronic exposure of 28 days. In absence of significant variations of parent compounds in milk, they observed a strong increase of PAH metabolites in milk, especially 1-OH-pyrene and 2-OH-fluorene (until 1.6% of the initial amount of the corresponding parent compound). This result suggests that soil-bound PAHs are available and submit a notable metabolism after their extraction during the digestive transfer. Costera (2007) compared the transfer of PAHs to milk and urine in lactating goats receiving either PAH contaminated hay or PAH contaminated soil. Concentrations of hydroxy-metabolites of PAHs were found to significantly increase in milk and in urine, but very similar kinetics were found for hay and soil.

4.3. TRANSFER PATTERNS

These results show that the transfer of POPs is related on the hydrophobicity and the metabolic susceptibility of the compounds. Previous paragraphs are focused on two families of classically separated POPs, the chlorinated compounds (PCDD/Fs and PCBs considered as persistent and bioaccumulable in livestock products) and the non-chlorinated PAH considered as very largely metabolized within the animal organism. Few are known about PBDEs but it seems that brominated POPs would be transferred in similar ways than chlorinated compounds. Data given above are in agreement with this global vision since the transfer rate coefficients are higher for the halogened families than for PAHs. Nevertheless large differences within the same family are observed, mainly for halogened POPs: the transfer rate varies from 5% to 90% for PCBs, and from 1% to 40% for PCDD/Fs. The relation between hydrophobicity of the compounds and their transfer to milk does not sufficiently explain the behaviour of all the indivi-dual compounds. Among PCB and PCDD/F which are known to be metabolized, the observed transfer rates are lower than expected when considering their level of hydrophobicity. These questions have also been raised by authors evaluating pollutant transfer in the aquatic (Baussant et al., 2001; Dearden, 2002) or terrestrial (Fries et al., 2002) food chain. The apparition of another degradation processes can be expected. The mechanisms responsible for these marked differences between compounds would be the *in vivo* biotransformation and/or absorption of the compounds. The ruminant animal seems to adapt to an increase in PAH exposure by an activation of its capacity of biotransformation, the contents of native molecules in milk increasing only slightly, whereas the OH-metabolites increase in higher amounts. These metabolites are much less hydrophobic than the parent molecules and can therefore be excreted through urine.

5. Conclusion

Dairy ruminants can be exposed to chlorinated POPs mainly by ingestion of con-taminated soil although other sources can also cause an exposure. Brominated POPs as PBDEs can enter into the food chain by the aquatic way and organic matter, which can get into contact to livestock systems. Finally, exposure of animals to PAHs can be due to contaminated soil or fodder as both matrixes can be heavily enriched in such com-pounds.

In the case of exposure, the transfer risk is elevated for chlorinated compounds as PCDD/Fs and PCBs. Few is known about the transfer of PBDEs but reported con-centrations in milk let assume transfer abilities similar to chlorinated POPs. PAHs are mainly transferred in metabolised forms and their impact on human health should be clarified.

References

Abramowicz, D., 1990, Aerobic and anaerobic biodegradation of PCBs: a review, *Criti. Rev. Biotechnol.*, 10: 241–251.

Arthur, M.F., Frea, J.I., 1989, 2,3,7,8TCDD: aspects of its important properties and its potential biodegradation in soils, *J. Environ. Quality*, 18: 1–12.

Bakker, D.J., Vries, W., 1996, Manual for calculating critical loads of persistent organic pollutants for soil and surface water. Energy Research and Process Innovation. TNO Institute of Environmental Sciences. Report R96/509, Delft, The Nederlands.

Bakker, M.I., Casado, B., Koerselman, J.W., Tolls, J., Kollöffel, C., 2000, Polycyclic aromatic hydrocarbons in soil and plant samples from the vicinity of an oil refinery. *Sci. Total Environ.* 263: 91–100.

Bakker, M.I., Tolls, J., Kollöffel, C., 2001, Deposition of atmospheric semivolatile organic compound to vegetation. American Chemical Society, Chapter 16, 218–236.

Baussant, T., Sanni, S., Skadsheim, A., Jonsson, G., Borseth, JF., Gaudebert, B., 2001, Bioaccumulation of polycyclic aromatic compounds: 2 Modeling bioaccumulation in marine organisms chronicallyexposed todispersed oil, *Environ. Toxicol. Chem.*, 20: 1185–1195.

Betts, K.S. 2001, Rapidly rising PBDE levels in North America. Environmental Science & Technology, Science News – 7 December 2001.

Beurskens, J.E.M., Toussaint, M., de Wolf, J., Van der Steen, J.M.D., Slot, P.C., Commandeur, L.C.M., Parsons, J.R., 1995, Dehalogenation of chlorinated dioxins by an anaerobic microbial consortium from sediment, *Environ. Toxicol. Chem.*, 14: 939–943.

Blais, J.M., Kimpe, L.E., Backus, S., Comba, M., Schindler, D.W., 2003, Assessment and characterization of PCB near a hazardous waste incinerator: analysis of vegetation, snow and sediments, *Environ. Toxicol. Chem.*, 22: 126–133.

Bocio, A., Llobet, J.M., Domingo, J.L., Corbella, J., Teixidoa, A., Casas, C., 2003, Polybrominated Diphenyl Esters (PBDEs) in foodstuffs: Human Exposure through the Diet, *J. Agric. Food Chem.* 51: 3191–3195.

Bonten, L., Grotenhuis, T.C., Rulkens, W.H., 1999, Enhancement of PAH biodegradation in soil physicochemical pre-treatment, *Chemosphere*, 38: 3627–3636.

Cerneglia, C.E., 1992, Biodegradation of PAH, *Biodegradation*, 3: 351–368.

Costera, A., Feidt, C., Marchand, P., Le Bizec, B., Rychen, G., 2006, PCDD/F and PCB transfer to milk in goats exposed to long-term intake of contaminated hay, *Chemosphere*, 64: 650–657.

Costera, A. 2007, Transfert des POP du fourrage au lait. Thèse INPL, 140 p., june 2007.

Dearden, J.C., 2002, Prediction of Environmental Toxicity and Fate Using Quantitative Structure-Activity Relationships (QSARs), *J. Braz. Chem. Soc.*, 13: 754–762.

Domingo, J.L., 2006, Polychlorinated Diphenyl Esters (PCDEs): Environmental levels, toxicity and human exposure. A review of the published literature, *Environ. Int.*, 32: 121–127.

Dougherty, E.J., McPeter, A.L., Ovecash, M.R., Carbonell, R.G., 1993, Theoretical analysis of a method for in situ decontamination of soil containing 2,3,7,8 TCDD, *Environ. Sci. Technol.*, 27: 505–515.

Edwards, N.T., 1983, Polycyclic aromatic hydrocarbons (PAH's) in the terrestrial environment. A review, *J. Environ. Quality*, 12: 427–441.

Eitzer, B.D., 1995. Polychlorinated dibenzo-p-dioxins and dibenzofurans in raw milk samples from farms located near a new resource recovery incinerator, *Chemosphere*, 30: 1237–1248.

Fängström, B., Athanasiadou, M., Athanassiadis, I., Bignert, A., Grandjean, Ph., Weihe, P., Bergman, A., 2005, Polybrominated diphenyl esters and traditional organochlorine pollutants in fulmars (*Fulmarus glacialis*) from the Faröe Islands, *Chemosphere*, 60: 836–843.

Firestone, D., Clower, M. Jr, Borsetti, A.P., Teske, R.H., Long, P.E., 1979, Polychlorodibenzo-p-dioxin and pentachlorophenol residues in milk and blood of cows fed technical pentachlorophenol, *J. Agric. Food Chem.*, 27: 1171–1177.

Focant, J.F., Pirard, C., Massard, A.C., de Pauw, E., 2003, Survey of commercial pasteurised cows' milk in Wallonia (Belgium) for the occurrence of polychlorinated dibenzo-*p*-dioxins, dibenzofurans and coplanar polychlorinated biphenyls, *Chemosphere*, 52: 725–733.

Fries, G.F., 1982, Potential polychlorinated biphenyl residues in animal product from application of contaminated sewage sludge to land, *J. Anim. Sci.*, 73: 1639–1650.

Fries, G.F., Paustenbauch, D.J. 1990, Evaluation of potential transmission of 2,3,7,8TCDD incinerator transmissions to humans via food, *Toxicol.Environ. Health*, 29: 1–43.

Fries, G.F., Paustenbach, D.J., Mather, D.B., Luksemburg, W.J., 1999, A congener specific evaluation of transfer of chlorinated dibenzo-*p*-dioxins and dibenzofurans to milk of cows following ingestion of pentachlorophenol-treated wood, *Environ. Sci. Technol.*, 33: 1165–1170.

Fries, G.F., Paustenbach, D.J., Luksemburg, W.J., 2002, Complete mass balance of dietary PCDD/F in dairy cattle and characterisation of the apparent synthesis of hepta-and octachlorodioxins, *J. Agric. Food Chem.*, 50: 4226–4231.

Gibson, D.T., Subramanian, V., 1984, Microbial degradation of aromatic hydrocarbons. In: Microbial degradation of organic compounds, D.T. Gibson (Ed). Marcel Dekker, New York, 525 p

Grova, N., Feidt, C., Crépineau, C., Laurent, C., Lafargue, P.E., Hachimi, A., Rychen, G., 2002a, Detection of polycyclic aromatic hydrocarbon levels in milk collected near potential contamination sources, *J. Agric. Food Chem.*, 2002, 50: 4640–4642.

Grova, N., Feidt, C., Laurent, C., Rychen, G., 2002b, Milk, urine and faeces excretion kinetics in lactating goats after an oral administration of [14C] polycyclic aromatic hydrocarbons, *Int. Dairy J.*, 12: 1025–1031.

Grova, N., Rychen, G., Monteau, F., Le Bizec, B., Feidt, C., 2006, Effect of oral exposure to polycyclic aromatic hydrocarbons on goat's milk contamination, *Agron. Sust. Dev.*, 26: 195–199.

Habe, H., Chung, J.S., Lee, J.H., Kasuga, K., Yoshida, T., Nojiri, H., Omori, T., 2001, Degradation of chlorinated dibenzofurans and dibenzo-p-dioxins by two types of bacteria having angular dioxygenases with different features, *Appl. Environ. Microbiol.*, 67: 3610–3617.

Habe, H., Die, K., Yotsumoto, M., Tsuji, H., Yoshida, T., Nojiri, H., Omori, T., 2002, Degradation characteristics of a dibenzofuran-degrader Terrabacter sp. Strain DBF63 toward chlorinated dioxins in soil, *Chemosphere*, 48: 201–207.

Harrison, N., de Gem, M.G., Starting, J.R., Wright, C., Kelly, M., Rose, M., 1996, PCDDs and PCDFs in milk from farms in Derbyshire, U.K, *Chemosphere*, 32: 453–460.

Hassanin, A., Breivik, K., Meijer, S., Livsteinnes, E., Garetho, Th., Andkevinc, J., 2004, PBDEs in European background soils: levels and factors controlling their distribution, *Environ. Sci. Technol.*, 38: 738–745.

Healy, G.B., 1968, Ingestion of soil by dairy cows, New Zealand, *J. Agric. Res.*, 11: 487–499.

Hebert, V.R., Miller, G.C., 1990, Deph dependence of direct and indirect photolysis on surface soil, *J. Agric. Food Chem.*, 38: 913–918.

Helling, C.S., Isensee, A.R., Woolson, E.A., Enso, P.J.D., Jones, G.E., Plimmer, J.R., Kearney, P.C., 1973, Chlorodioxins in pesticides, soils and plants, *J. Environ. Quality*, 2: 171–178.

Henner, P., 2000, Phytorémédiation appliquée au traitement des sols contaminés par des hydrocarbures aromatiques polycycliques. These de doctorat de l'INPL, Nancy, 187 p.

Hippelein, M., Kaupp, H., Dörr, G., McLachlan, M.S., Hutzinger, O., 1996, Baseline contamination assessment for a new resource recovery facility in Germany part II: atmospheric concentrations of PCDD/F, *Chemosphere*, 32: 1605–1616.

Howsam, M., Jones, K.C., Ineson, P., 2000, PAHs associated with the leaves of three deciduous tree species. I. Concentrations and profiles, *Environ. Poll.*, 108: 413–424.

Jan, J., Zupancic-Kralj, L., Kralj, B., Marsel, J., 1994, The influence of exposure time and transportation routes on the pattern of organochlorines in plants from a polluted region, *Chemosphere*, 29: 1603–1610.

Jarrige., R. 1988. Alimentation des bovins, ovins et caprins. Ed. INRA, France, 471 p.

Jarrige, R., Ruckebusch, Y., Demarquilly, C., Fonce, M.H., Journet M. 1995, Nutrition des ruminants domestiques – ingestion et digestion, Ed. INRA Paris, 920 p.

Jones, KC., Stratford, JA., Tidridge, P., Waterhous, KS., Johnston, AE., 1989, Polynuclear aromatic hydrocarbons in an agricultural soil: long-term changes in profile distribution, *Environ. Poll.*, 56: 337–351.

Juhasz, A.L., Naidu, R., 2000, Bioremediation of high molecular weight polycyclic aromatic hydrocarbons: a review of the microbial degradation of benzo[a]pyrene, *Int. Biodeter. Biodegrad.*, 45: 57–88.

Kakareka, S.V., 2002, Sources of persistent organic pollutants emissions on the territory of Belarus, *Atmos. Environ.*, 36: 1407–1419.

Kästner, M., Mahro, B., 1996, Microbial degradation of polycyclic aromatic hydrocarbons in soils affected by the organic matrix of compost, *Appl. Microbiol. Biotechnol.*, 44: 668–675.

Kieaitwong, S., Nguyen, L.V., Herbert, V.R., Hackett, M., Miller, G.C., 1990, Photolysis of chlorinated dioxins in organic solvents and soils, *Environ. Sci. Technol.*, 24: 1575–1580.

Kipopoulou, A.M., Manoli, E., Samara, C., 1999, Bioconcentration of polycyclic aromatic hydrocarbons in vegetables grown in an industrial area, *Environ. Poll.*, 106: 369–380.

Kirby, D.R., Stuth, J.W., 1980, Soil ingestion rates of steers following brush management in Central Texas, *J. Range Manag.*, 33: 207–209.

Kiviranta, H., Ovaskainen, M.L., Vartiainen, T., 2004, Merket basket study on dietary intake of PCDD/Fs, PCBs and PBDEs in Finland, *Environ. Int.*, 30: 923–932.

Krauss, M., Wilcke, W., 2003, Polychlorinated naphthalenes in urban soils: analysis, concentrations and relation to other persistent organic pollutants, *Environ. Poll.*, 122: 75–89.

Krokos, F., Creaser, C.S., Wright, C., Startin, J.R., 1996, Levels of selected *ortho* and non-*ortho* polychlorinated biphenyls in UK retail milk, *Chemosphere*, 32: 667–673.

Lapole, D., Rychen, G., Monteau, F., Le Bizec, B., Feidt, C., 2007, Transfer of phenanthrene, pyrene and benzo[a]pyrene and their major metabolites to blood, milk and urine, *J. Dairy Sci.* 90: 2624–2629.

Laurent, C., Feidt, C., Laurent, F., 2005, Etat de l'art sur les transferts de polluants organiques et métalliques du sol vers l'animal, Ademe, pp. 1–216, EDP Sciences, Les Ulis.

Law, R.J., Allchin, C.R., de Boer, J., Covaci, A., Herzke, D., Lepom, P., Morris, S., Tronczynski, J., de Wit, C.A., 2006, Levels and trends of brominated flame retardants in the European environment, *Chemosphere*, 64: 187–208.

Lichfouse, E., Albercht, P., Behar, F., Hayes, JM., 1994, A molecular and isotopic study of the organic matter from Paris Basin, France, *Geochim. Cosmochim. Acta*, 58: 209–221.

Lichfouse, E, Budzinski, H, Garrigues, PH, Eglinton, TI, 1997, Ancient polycyclic aromatic hydrocarbons in modern soils: ^{13}C, ^{14}C and biomarker evidence, *Org. Geochem.*, 26: 353–359.

Lohmann, R., Jones, K.C., 1998, Dioxines and furans in air and deposition: a review of levels, behaviour and process. *Sci. Total Environ.* 219: 53–81.

Lohman, K., Seigneur, C., 2001, Atmospheric fate and transport of dioxins: local impacts, *Chemosphere*, 45: 161–171.

Lutz, S., Feidt, C., Monteau, F., Rychen, G., Le Bizec, B., Jurjanz, S., 2006, Effect of exposure to soil_bound polycyclic aromatic hydrocarbons on milk contaminations of parents compounds and their monohydroxylated metabolites, *J. Agric. Food Chem.*, 54: 263–268.

Mamontova, E.A., Tarasova, E.N., Momonto, A.A., Kusmin, M.I., McLachlan, M.S., Khomutova, M. Iu., 2007, The influence of soil contamination on the concentrations of PCBs in milk in Siberia, *Chemosphere*, 9: S71–S78.

Mayland, H.F., Florence, A.R., Rosenau, R.C., Lazar, V.A., Turner, H.A., 1975, Soil ingestion by cattle on semiarid range as reflected by titanium analysis of feed, *J. Range Manag.*, 28: 448–452.

McLachlan, M.S., Thoma, H., Reissinger, M., Hutzinge, O., 1990, PCDD/F in an agricultural food chain. Part1: PCDD/F mass balance of a lactating cow, *Chemosphere*, 20: 1013–1020.

McLachlan, M.S., 1993, Mass balance of polychlorinated biphenyls and other organochlorine compounds in a lactating cow, *J. Agric. Food Chem.*, 46: 1166–1172.

McLachlan, M.S., 1999, Framework for the interpretation of measurements of SOCs in plants, *Environ. Sci. Technol.*, 33: 1799–1804.

McPeters, A.L., Overcash, M.R., 1993, Demonstration of photodegradation by sunlight of 2,3,7,8-TCCD in 6 cm soil columns, *Chemosphere*, 20: 1013–1020.

Meneses, M., Schumacher, M., Domingo, J.L., 2002, A design of two simple models to predict PCDD/F concentrations in vegetation and soils, *Chemosphere*, 46: 1393–1402.

Mhiri, C., de Marsac, N.T., 1997, Réhabilitation par les microorganismes de sites contenant du pyralène: problématique et perspectives d'étude, *Bulletin de l'Institut Pasteur*, 95: 3–28.

Miller, G.C., Hebert, V.R., Mille, M.J., Mitzel, R., Zepp, R.G., 1989, Photolysis of octachlorodibenzo-p-dioxin on soils: production of 2,3,7,8 TCDD, *Chemosphere*, 18: 1265–1274.

Moore, J.W., Ramworthy, S., 1984, Organic chemicals in natural waters. Springer-Verlag, New York.

Müller, J.F., Hawker, D.W., McLachlan, M.S., Connel, D.W., 2001, PAHS, PCDD/Fs, PCBs and HCB in leaves from Brisbane, Australia, *Chemosphere*, 43: 507–515.

Olling, M., Derks, H.J.G.M, Berende, P.L.M, Liem, A.K.D., Jong, A.P.J.M., 1991, Toxicokinetics of eight ^{13}C-labelled polychlorinated dibenzo-p-dioxins and − furans in lactating cows, *Chemosphere*, 23: 1377–1385.

Paustenbach, D.J., Weining, R.J., Lau, V., Harrington, N.W., Rennix, D.K., Parsons, A.H., 1992, Recent developments on the hazards posed by 2,3,7,8 TCDD in soil: implications for setting risk-based cleanup levels at residential and industrial sites, *J. Toxicol. Environ. Health*, 36: 103–149.

Ramos, L., Eljarrat, E., Hernandez, L.M., Alonso, L., Rivera, J., Gonzalez, M.J., 1997, Levels of PCDDs and PCDFs in farm cow's milk located near potential contaminant sources in Asturias (Spain). Comparison with levels found in control, rural farms and commercial pasteurized cow's milks, *Chemosphere*, 35: 2167–2179.

Rappe, C., Nygren, M., Lindström, G., Buser, H.R., Blaser, O., Wüthrich, C., 1987, Polychlorinated dibenzofurans and dibenzo-*p*-dioxins and other chlorinated contaminants in cow milk from various locations in Switzerland, *Environ. Sci. Technol.*, 21: 964–970.

Rychen, G., Ducoulombier-Crépineau, C., Grova, N., Jurjanz, S., Feidt, C., 2005, Modalités et risques de transfert des polluants organiques persistants vers le lait, *Inra Productions Animales*, 18(5): 355–366.

Schiavon, 1998, Studies of the leaching of atrazine, of its chlorinated derivatives, and hydroxyatrazine from soil using 14C ring-labelled compounds under outdoor conditions, *Ecotox. Environ. Safety*, 15: 46–54.

Schmid, P., Schlatter, C., 1992, Polychlorinated dinebzo-p-dioxins (PCDDs) and polyclorinated dibenzofurans (PCDFs) in cow's milk from Switzerland, *Chemosphere*, 8: 1013–1030.

Schumacher, M., Agramunt, M.C., Bocio, A., Domingo, J.L., de Kok, H.A.M., 2002, Annual variation in the levels of metals and PCDD/PCDFs in soil and herbage samples collected near a cement plant, *Environ. Int.*, 29: 415–421.

Schumacher, M., Nadal, M., Domingo, J.L., 2004, Levels of PCDD/Fs, PCBs, and PCNs in soils and vegetation in an area with chemical and petrochemical industries, *Environ. Sci. Technol.*, 38: 1960–1969.

Sewart, A., Jones, K.C., 1996, 'A survey of PCB congeners in UK cows' milk, *Chemosphere*, 32, 2481–2492.

Sierra, I., Valera, J.L., Marina, M.L., Laborda, F., 2003, Study of the bioremediation process of polychlorinated byphenyls in liquid medium and soil by a new isolated aerobic bacterium (Janibacter sp.), *Chemosphere*, 6: 609–618.

Simonich, S.L., Hites, R.A., 1994, Vegetation-atmosphere partitioning of polycyclic aromatic hydrocarbons, *Environ. Sci. Technol.*, 28: 939–943.

Sinkkonen, S., Paasivirta, J., 2000, Degradation half-life times of PCDDs, PCDFs and PCBs for environmental fate modelling, *Chemosphere* 40: 943–949.

Smith, K.E.C., Thomas, G.O., Jones, K.C., 2001, Seasonal and species differences in the air-pasture transfer of PAHs, *Environ. Sci. Technol.*, 35: 2156–2165.

Stalikas, C.D., Chaidou, C.I., Philidis, G.A., 1997, Enrichment of PAHs and heavy metals in soils in the vicinity of the lignite-fired power plants of West Macedonia (Greece), *Sci. Total Environ.*, 204: 135–146.

Steven, J.L., Gerbec, E.N., 1988, Dioxin in agricultural food chain, *Risk Anal.*, 8: 329–335.

Straube, W.L., Jones-Meehan, J., Pritchard, P.H., Jones, W.R., 1999, Bench-scale optimisation of bioaugmentation for the treatment of soils contaminated with molecular weight polyaromatic hydrocarbons, *Resour. Conserv. Recycl.*, 27: 27–37.

Teil, M.J., Blanchard, M., Chevreuil, M., 2004, Atmospheric deposition of organochlorines (PCBs and pesticides) in northern France, *Chemosphere*, 55: 501–514.

Thomas, G.O., Smith, K.E.C., Sweetman, A.J., Jones, K.C., 1998, Further studies of the air–pasture transfer of polychlorinated biphenyls, *Environ. Poll.*, 102: 11–128.

Thomas, G.O., Sweetman, A.J., Jones, K.C., 1999, Metabolism and body-burden of PCBs in lactating dairy cows, *Chemosphere*, 39: 1533–1544.

Thornton, I., Abraham, P., 1983, Soil ingestion – A major pathway of heavy metals into livestock grazing contaminated land, *Sci. Total Environ.*, 28: 287–294.

Tuinstra, L.G.M.Th., Roos, A.H., Berende, P.L.M., van Rhijn, J.A., Traag, W.A., Mengelers, M.J.B., 1992, Excretion of polychlorinated dibenzo-p-dioxins and -furans in milk of cows fed on dioxins in the dry period, *J. Agric. Food Chem.*, 40: 1772–1776.

Tysklind, M., Carey, A.E., Rappe, C., Miller, G.C., 1992, Photolysis of OCDF and OCDD on soil, *Organohalogen Comp.*, 8: 293–296.

United States Environmental Protection Agency (US EPA), 2000, Draft exposure and human health reassessement of 2,3,7,8-TCDD and related compounds, National Academy Sciences (NAS) http://cfpub.epa.gov september 2000.

United States Environmental Protection Agency (US EPA), 2003, Table of PCB species by CAS registry number, US EPA Region 5, www.epa,gov/toxteam/pcbid/, 5 November 2003.

Welsch-Pausch, K., McLachlan, M.S., Umlauf, G., 1995, Determination of the principal pathway of polychlorinated dibenzo-p-dioxins and dibenzofurans to *lolium multiflorum* (Welsh Ray Grass), *Environ. Sci. Technol.*, 29: 1090–1098.

Welsch-Pausch, K., McLachlan, M.S., 1998, Fate of airborne polychlorinated dibenzo-*p*-dioxins and dibenzofurans in an agricultural ecosystem, *Environ. Poll.*, 102: 129–137.

Wild, S.R., Jones, K.C., 1992, The polynuclear aromatic hydrocarbon (PAH) content of herbage from a long-term grass-land experiment, *Atmos. Environ.*, 26a(7): 1299–1307.

Wild, S.R., Jones, K.C., 1995, Polycyclic aromatic hydrocarbons in the United Kingdom environment: a preliminary source inventory and budget, *Environ. Poll.*, 88: 91–108.

Willett, L.B., Liu, T.T., Durst, H.I., Smith, K.L., Redman, D.R., 1987, Fundamental and applied toxicology, *Off. J. Soc. Toxicol.*, 9: 60–68.

Willett, K.L., Loerch, S.C., Willett, L.B., 1989, Effects of halogenated hydrocarbons on rumen microorganisms, *J. Vet. Diagn. Invest.*, 1, 120 p.

Wilson, S.C., Jones, K.C., 1993, Bioremediation of soil contaminated with polynuclear aromatic hydrocarbons (PAHs): a review, *Environ. Poll.*, 81: 229–249.

Wittich, R.M., 1998, Degradation of dioxin-like compounds by microorganisms, *Appl. Microbiol. Biotechnol.*, 49: 429–447.

Zurbier, M., Leijs, M., Schoeters, G., ten Tusscher, G., Koppe, J.G., 2006, Children's exposure to polybrominated diphenyl ethers, *Acta Paediatrica*, 95: 65–70.

Annexe: Figure 1

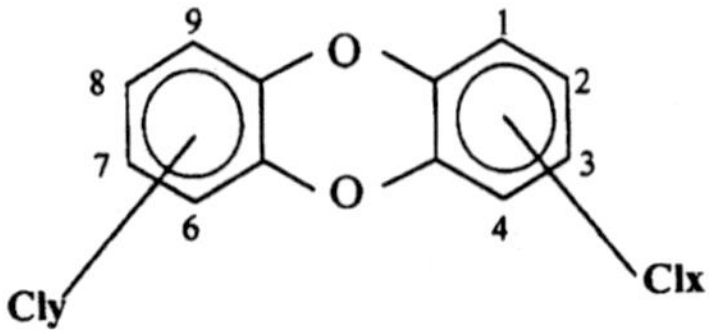

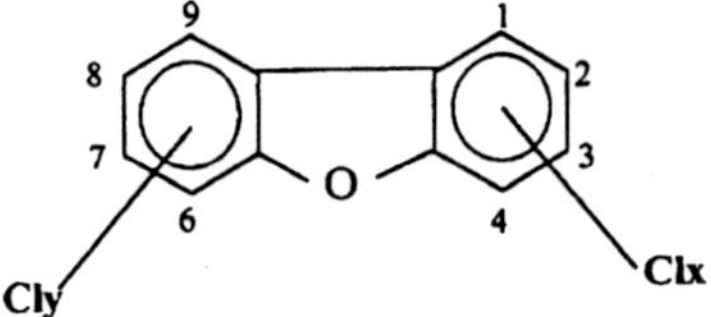

Figure 1a. Chemical structure of polychlorodibenzo-para-dioxins (PCDDs)

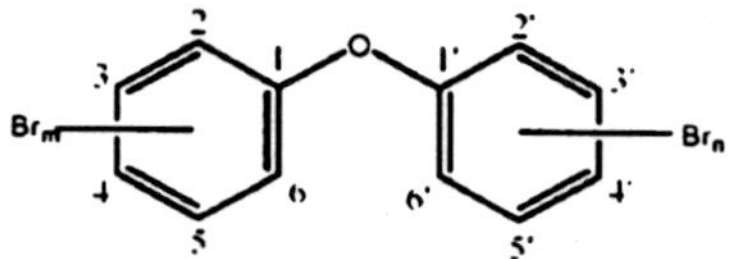

Figure 1b. Chemical structure of polychlorodibenzofurans (PCDFs)

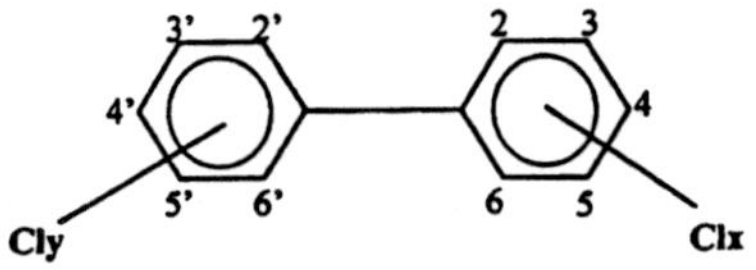

Figure 1c. Chemical structure of polychlorobiphenyls (PCBs)

Figure 1e. Chemical structure of polybromodiphenylethers (PBDEs)

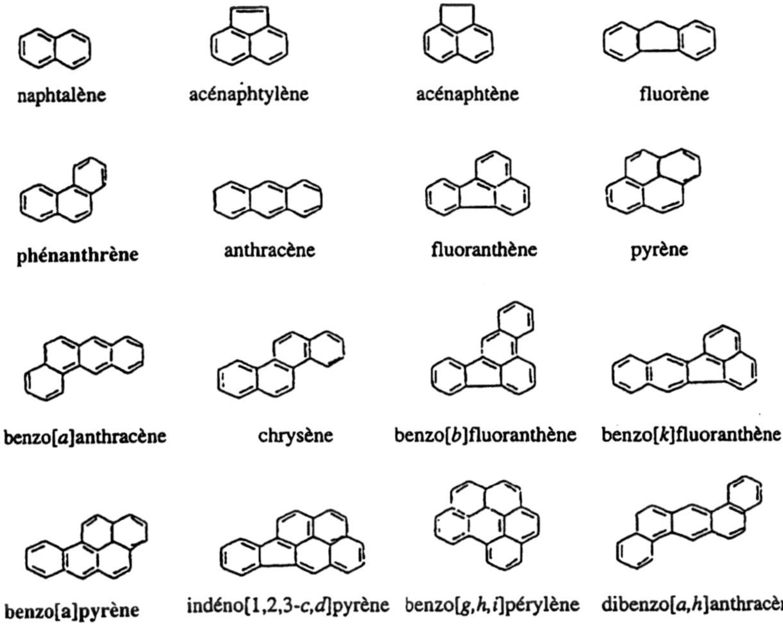

Figure 1d. Chemical structure of polycyclic aromatic hydrocarbons

HYDROTELLURIC AND INDUSTRIAL FLUOROSIS SURVEY IN THE DROMEDARY CAMEL IN THE SOUTH OF MOROCCO

EMILIE DIACONO, BERNARD FAYE
Département Environnement et Société, CIRAD, Montpellier, France

MOHAMMED BENGOUMI, MOHAMED KESSABI
Département de Pharmacie, Toxicologie et Biochimie, Institut Agronomique et Vétérinaire Hassan II, Rabat, Maroc

Abstract: Morocco is the first country producer of phosphate in the world with a real potential of contamination of the environment and individuals there living by fluorine either by phosphate deposits (hydrotelluric fluorosis) and phosphate manufacturing plants (industrial fluorosis). This survey was achieved on 86 dromedaries in a region of the Sahara (Boujdour and Laâyoune) characterized by the presence of phosphate. In addition, blood, soil, water and plant samples were collected for the dosage of fluorine that has been achieved by potentiometric method. The mean fluorine content was below 0.47 ppm, 513 ppm and 4.8 ppm in water, soil and plants respectively. The provinces of Boujdour and Laâyoune are unscathed zones opposite the sources of fluorine contamination, as water, vegetation and soil. The mean plasma fluorine concentration was below 0.06 ppm, thus, the camels of these regions seem therefore also free of fluorine chronic intoxication. However the increased values of fluorine levels in the soil, vegetables, and the plasma of camels in the region of Boujdour can let suppose that this area is close to a source of fluorine contamination. Indeed, the province of Boujdour is located unless 200 km of Boukraa where is situated a processing plant of phosphates. Thus, according orientation and the strength of the present dominant winds in the region of Boujdour, we can give out the hypothesis that by winds are brought in the region of Boujdour of the fluorine particles coming from the region of Boukraa. These winds carrying particles of fluorine eliminated by the factory and also by the extraction of soil particles by erosion.

This hypothesis can be verified by a survey establishing a gradient of pollution by fluorine cleared by the deposit or the processing plant of the phosphates considering the direction and the strength of the dominant winds in these regions.

Keywords: Dromedary, fluorine, soil, plasma, water, pollution

1. Introduction

Fluorosis is a chronic intoxication by fluorine. Plants, animals and human could be affected by this intoxication due to the presence of phosphates rocks in the soil (Kessabi et al., 1984). By erosion of soil, fluorine compounds in excess contaminate plants and water (Laatar et al., 2003). In this way, human and animals could be intoxicated by high fluorine level in water and plants (Clark et al., 1976).

Two kinds of fluorosis were described: hydrotelluric and industrial fluorosis. The first one is caused by high content of fluoride in the soil and water especially in phosphate rocks rich in fluoride. The second one is due to air emission of fluoride particle in the environment by phosphate plants and manufactories (Kessabi et al., 1984).

Morocco is the first producer and exporter of phosphate in the word with many factories producing phosphoric acid and derivates from phosphate rock. Fluorine contamination is very important in some areas producing phosphate with hydrotelluric fluorosis and industrial fluorosis in other areas near phosphate plants.

Fluorine is characterized its high chemical affinity to calcium and then to calcified tissues including bone and teeth. The main lesions due to fluorine intoxication are modification of color, structure and orientation of teeth, and also structure and texture of bones. Fluorosis has an important impact on animal health and welfare. The precocious grinding of teeth is responsible for the low production level of milk and meat and also precocious animal culling. Meanwhile, the severity level of intoxication is variable towards different factors. The major extrinsic factors are fluorine content in feed and water, exposition duration, feed characteristics (quantity, nature and solubility of fluorine compounds and phospho-calcic balance), season and stress play an important role for expression of this intoxication. In fact, severity of lesions increases with fluorine quantity ingested and with the duration of exposition and toxicity of fluorine compounds (Reddy Sriranga and Srikantia, 1971; Kessabi et al., 1984). Intrinsic factors include species (herbivore are more sensible), breed, age, sex, individual, stress and concomitant disease (parasitic and infectious diseases). Young animals (after weaning) are more sensitive to this intoxication compare to adults (Zouagui, 1973). In fact, fluorine toxicity on teeth is more efficient during formation of adult teething (Zouagui, 1973; Underwood and Suttle, 1999). Toxic action of fluorine is function of stage of teeth formation during intoxication (Abdennebi, 1982). That's why teeth lesions could be observed only on definitive teeth of adults whom were exposed to high levels of fluorine during teeth formation (Zouagui, 1973; Laatar et al., 2003). Bones lesions are observed when the maximal capacity of fluorine fixation by bones is exceeded (Laatar et al., 2003).

The main objective of this study is to evaluate hydrotelluric and industrial fluorosis in camels by measuring fluorine in plasma, soil, water and plants in the south of Morocco.

2. Materials and Methods

2.1. STUDY AREA

This study was performed in the south west of Morocco, in two provinces: Boujdour characterized by the presence of natural phosphate rock deposits and Laâyoune with an important phosphate manufactory.

2.2. ANIMALS

Eighty six camels were sampled for plasma (50 in Boujdour, 36 in Laâyoune). On average the camels sampled in the Boujdour province were 6.11 year olds, and 3.44 year olds in the Laâyoune province.

Four camels from Marrakech were sampled as control, because this area has no phosphate deposit or phosphate manufactory.

2.3. SURVEY AND SAMPLES

One inquest was fulfilled to nine camel's breeders. In order to collect general information on herd and to evaluate how the breeders perceive fluorosis symptoms and what is their attitude towards it.

In addition water and plants consumed by camels and soil were sampled in different suspected contaminated areas and also in free fluorosis, as control areas especially Rabat for water and soil samples and Marrakech for camel blood samples.

2.4. LABORATORY ANALYSES

Specific fluorine electrode was used to determinate the content of fluorine in each sample. Plant and soil samples were mineralized by using dry method in a muffle furnace at 550°C during 12 h and then solubilized in distilled water (Kessabi et al., 1984).

2.5. STATISTICAL ANALYSIS

Statistical comparisons between data were performed using t-Student and Mann-Whitney tests. Relationships between data were assessed by Kendall's method.

3. Results

3.1. FLUORINE CONCENTRATION IN WATER

The mean fluorine concentration in water was 0.134 ± 0.039 ppm in Boujdour and 0.171 ± 0.051 ppm in Laâyoune whereas control sample concentration was 0.031 ppm. There was no significant difference between these three areas. The maximum value was observed in Boujdour (0.47 ppm). These values are under the toxic limit of fluorine in water (1 ppm) (Figure 1).

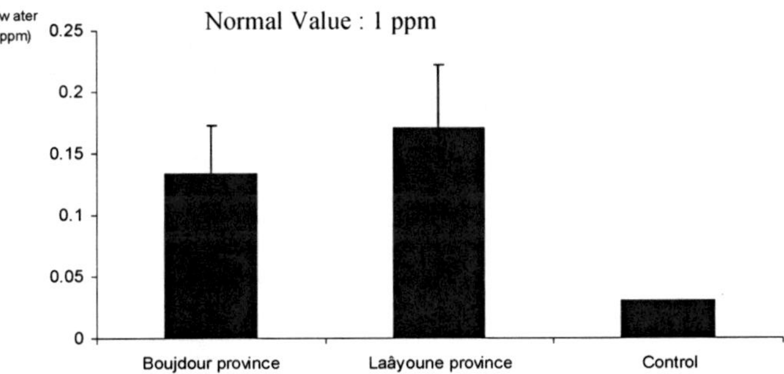

Figure 1. Mean fluorine concentration in water (±SE)

3.2. CONTENT OF FLUORINE IN SOIL

The average fluorine content in soil was 513 ± 25 ppm in Boujdour and 578 ± 49 ppm in the control area. No samples were taken in Laâyoune. There was no significant difference ($p > 0.05$) in suspected contaminated area (Boujdour) and control area. The higher content of fluorine in soil that was reported in Boujdour (601 ppm) is lower than the toxic limit of fluorine in soil ($1,000$ ppm) (Figure 2).

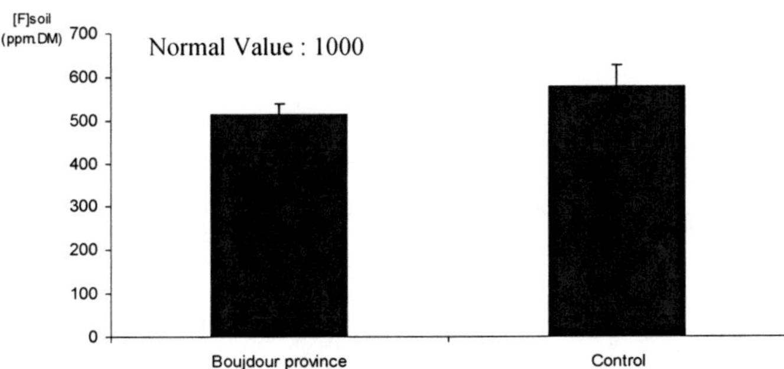

Figure 2. Average soil content of fluorine (±SE)

3.3. PLANT FLUORINE CONTENTS

Plant sample fluorine content was on average 4.80 ± 1.6 ppm and of 3.70 ± 0.09 ppm in Boujdour and Laâyoune respectively. The mean content of fluorine in controls samples was 1.49 ± 0.09 ppm. Only one sample from Boujdour had higher value (16.38 ppm) than the tolerable limit (15 ppm). No significant difference was observed between samples from Boujdour, Laâyoune and control areas (Figure 3).

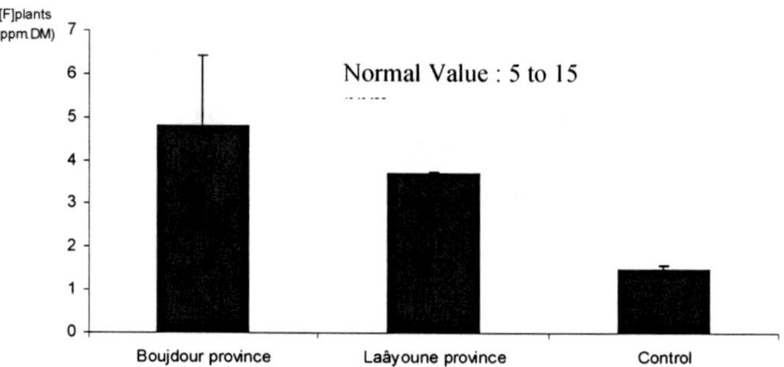

Figure 3. Mean fluorine content (±SE) in plants

3.4. PLASMA FLUORINE CONCENTRATION

Average camel plasma fluorine concentration was 0.064 ± 0.004 ppm in Boujdour and 0.045 ± 0.001 ppm in Laâyoune whereas it was 0.042 ± 0.001 ppm in control areas. The camel plasma fluorine concentration in Boujdour was significantly (p < 0.001) higher than in Laâyoune and control area. There was no significant difference (p = 0.4394) between plasma fluorine concentration of camels from Laâyoune and control area. In Boujdour, the plasma fluorine concentration was significantly higher in less than 1 year old camels than older camels. Sex and physiological status seemed to do not have any significant effect on plasma fluorine concentration (Figure 4).

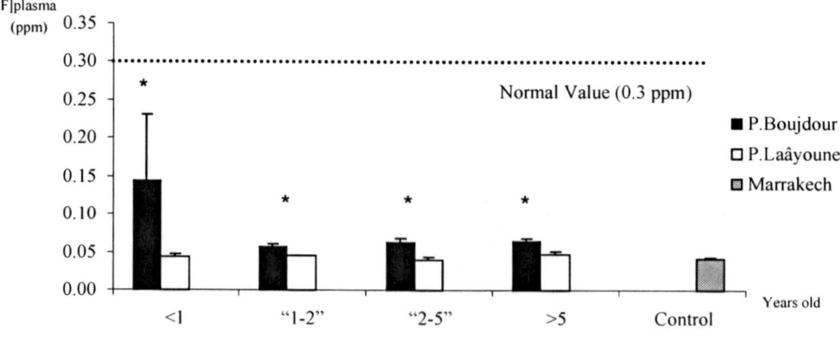

Figure 4. Effect of age on the mean camel plasma fluorine concentration (M ± SE)

3.5. RELATIONSHIP BETWEEN SOIL WATER, PLANTS AND BLOOD FLUORINE CONTENTS

In Boujdour, there was no significant interrelationship between fluorine content in soil and the plants ($p = 0.83$), between soil and water ($p = 0.74$) and between plants and water ($p = 0.52$) and in Laâyoune plants and water ($p = 0.60$).

4. Discussion

Camel fluorosis was already reported in Egypt by Karram et al. (1989). The current study is the first investigation on fluorosis survey in the dromedary camel in Morocco. For that, plants, soil and water was sampled in different areas of south Morocco, associated to samples of plasma of camels rearing in these areas.

Fluorine contents in soil samples are not in excess and under the limit value. Moreover, the control soil samples had on average a higher content in fluorine than the Boujdour soil samples (578 ± 49 ppm and 513 ± 25 ppm, respectively). These results are lower than those reported in the bibliography in other area of Morocco. In fact, in phosphate Northern areas, fluorine content of soil varied from 1,365 ppm (Abdennebi, 1982) to 4,545 ppm (Bettar, 1984). Fluorine content in plants has to be less than the toxic limit 15 ppm. In our study, fluorine content in plant samples was always under this limit in the two areas of study. Although, plants sampled in Boujdour province seemed to contain more fluorine (4.80 ± 1.62 ppm) than plants from Laâyoune province (3.71 ± 0.09 ppm). There were no significant differences between plants fluorine content in Boujdour, Laâyoune and control area. Water samples from the two areas of study have showed fluorine concentration on average under the toxic limit (1 ppm). Indeed, in Boujdour and Laâyoune, fluorine contents of water samples were: 0.13 ± 0.04 ppm and 0.17 ± 0.05 ppm respectively. These values were not different from fluorine content of control water samples (0.03 ppm).

According to the environmental factors analyzed in this case study, it appears that areas of sampling were not exposed to a fluorine contamination. Probably, these areas, especially Boujdour province have to face to a potential source of contamination by fluorine. Sampled areas in Boujdour province seem to be more exposed to an industrial fluorosis than those in Laâyoune, because they were nearest from Boukraa (place of phosphate manufactory) and more exposed to major wind coming from Boukraa. That's why, samples of water, soil and plants seem to contain more fluorine than in Laâyoune.

These results are in agreement with the bibliographic references on this subject. In fact, Alary (1970) and Assimi (1980) have reported that fluorine content in plant decreased with increase of distance from phosphate manufacturing, and also that force and orientation of wind is determinate factors of fluorine particle deposits.

In both regions, Boujdour and Laâyoune, there were no excess of fluorine in camel's plasma (under the limit value of 0.3 ppm). However, plasma fluorine concentration in camels from Boujdour was significantly higher than that from Laâyoune and control camels ($p < 0.001$). This result could be explained by the mean age of the camels sampled. In fact, camels from Boujdour were older on average than camels from Laâyoune (6.11 and 3.44 years old respectively). According to literature, fluorine concentration in plasma increases with the age of animal, after saturation of bones and teeth with this element. Underwood and Suttle (1999) showed that fluorine fixation is more important in bones and teeth during their growing (in young animal) than in mineralized bones (adult animal). Furthermore, Kessabi et al. (1984) and Cronin et al.

(2000) reported that fluorine concentration in blood reflected fluorine content in alimentation. Indeed, this could still explain the difference in fluorine plasma concentration of camels between regions, because of higher water and plants fluorine content in Boujdour.

Moreover, in Boujdour, significant difference in fluorine concentration in plasma according to the age of camels was observed ($p < 0.05$). Nevertheless this result was not in agreement with references because young camels (aged less than 1 year old) have significant more fluorine in plasma than older camels. This difference could not be explained here.

5. Conclusion

The only use of biochemical analysis was not enough to state the levels of fluorine contamination in the study area. In fact, according the environmental results (plants, water and soil), Boujdour and Laâyoune provinces seemed to be not affected by fluorine contamination in spite of the proximity of a potential source of phosphate. Concerning analysis on animals, plasma samples were sufficient, bones samples, clinical diagnosis and epidemiological data seem to be necessary to assess the real fluorosis survey in camels in the Southern Morocco.

Further studies should be performed with spatial analysis around pollution source near Boukraa (Phosphate manufacturing) taking in account climatic factors as wind (orientation, intensity, ...) and seasonal data and including other ruminants.

References

Abdennebi, E.H., 1982, Le Darmous: étude épidémiologique clinique et conséquences économiques. Thèse Méd. Vét., IAV Hassan II, Rabat, Morocco.

Alary, J., 1970, Contribution à l'étude des pollutions atmosphériques d'origine industrielle par les dérivés minéraux du fluor. Thèse Doct. Pharm., Paris, France.

Assimi, B., 1980, Contribution à l'étude de la fluorose industrielle dans la région de Safi. Thèse Méd. Vét., IAV Hassan II, Rabat, Morocco.

Bettar, A., 1984, Contribution à l'étude épidémiologique et biologique de la fluorose hydrotellurique (Darmous) en région phosphatière. Thèse Méd. Vét., IAV Hassan II, Rabat, Morocco.

Clark, R.G., Hunter, A.C., Stewart, D.J., 1976, Deaths in cattle suggestive of subacute fluorine poisoning following the ingestion of superphosphate, *New Zealand Vet. J.*, 24: 193–194.

Cronin, S.J., Manoharan, V., Hedley, M.J., Loganathan, P., 2000, Fluoride: a review of its fate, bioavailability and risks of fluorosis in grazed-pasture systems in New Zealand, *New Zealand J. Agric. Res.*, 43: 295–321.

Karram, M.H., Mottelib, A.A., Nafie, T.H.S., Sayed, A.S., 1989, Clinical and biochemical studies in chronic fluorosis and sulphurosis in camels, *Assiut Vet. Med. J.*, 21: 41–48.

Kessabi, M., Assimi, B., Braun, J.P., 1984, The effects of fluoride on animals and plants in the South Safi zone, *Science Total Environ.*, 38: 63–68.

Laatar, A., Mrabet, D., Zakraoui, L., 2003, La fluorose en Afrique subsaharienne, *Revue du rhumatisme*, 70: 178–182.

Reddy Sriranga, G., Srikantia, S.G., 1971, Effect of dietary calcium, vitamin C and protein in development of experimental skeletal fluorosis. 1. Growth, serum chemistry, and changes in composition, and radiological appearance of bones, *Metabolism*, 20(7): 642–649.

Underwood, E.J., Suttle, M.F., 1999, The mineral nutrition of livestock, 3rd ed. CABI Publishing, England.

Zouagui, H., 1973, Contribution à l'étude de la fluorose chez les grands et les petits ruminants au Maroc. Thèse Méd. Vét., Toulouse, France.

COPPER, ZINC, CADMIUM, AND LEAD IN SHEEP GRAZING IN NORTH JORDAN

MU'TAZ AL-ALAWI
Mu'tah University, P.O. Box 3, Karak 61710, Jordan

Abstract: The aim of this study is to investigate the levels of Cu, Zn, Cd and Pb in sheep in connection to the area of housing and grazing. In this study sheep tissue samples were examined in respect to the concentration of copper (Cu), zinc (Zn), cadmium (Cd) and lead (Pb). The samples were taken from sheep 3–7 years old and from three areas of Northern Jordan. Sheep tissue concentration of Cu in area A is high, while in the other two areas it is at normal levels. This high concentration must be attributed mainly to the spraying with copper compounds at fruit-tree cultures. It is therefore wise to avoid grazing at these areas during the season of spraying. Zinc concentration in sheep liver and kidney is considered to be at low levels. Therefore, addition of Zn in feed is recommended. Cadmium concentration is at desirable levels. Finally, lead concentration is at about the same level in all three areas and these values considered desirable.

Keywords: Sheep, Jordan, pollution, tissue, heavy metals

1. Introduction

The industrial evolution, the intense use of raw material and the agricultural technology have all some how improved man's living while simultaneously have polluted the natural environment with consequences on his health. Some of the more toxic elements are heavy metals which are absorbed through food by people not involved with this by profession (Schuhmacher et al., 1991). It is also necessary to point out the fact that some of these heavy metals (Pb, Cd) are implicated in causing cancer or mutations (Jacobson and Turner, 1980) while at the same time man's load of these elements in comparison to the last century has quadrupled (Ellinder and Kjelstrom, 1977).

The aim of this study is to investigate the levels of Cu, Zn, Cd and Pb in sheep in connection to the area of housing and grazing. The reason for choosing sheep for this investigation is that these animals live and feed for the greatest part of the year outdoors and, thus, the concentration of heavy metals in this species reflects the environmental load more accurately than in other productive animals.

2. Materials and Methods

In this study sheep tissue samples (70 liver and 70 kidney samples) were examined in respect to the concentration of copper (Cu), zinc (Zn), cadmium (Cd) and lead (Pb). The samples were taken from sheep 3–7 years old and from three areas of Northern Jordan.

Area A: this area had large areas of fruit tree cultures and vegetable cultures where many incidents of copper poisoning occurred.

Area B: this area was characterized by many fruit-tree cultures but, due to the presence of sulphur radicals in the soil, was considered poor in copper.

Area C: this area was mostly highlands and human intervention on its ecosystem was minimal.

The samples were stored at −20°C until examination. For the measurements, 20 g of tissues were homogenized and wet digested with nitric and perchloric acid (7:3 ratio). For the Cu and Zn assay, direct measurement by Atomic Absorption Spectrometry

(A.A.S.) was performed. For Pb and Cd determinations, pH values were adjusted to 4, complexes with ammonium pyrrolidinum dithiocarbonate were formed and extracted in methyl-isobutyl-ketone. Finally, the organic layer was aspirated in flame AAS at appropriate wave lengths with cathode lamps (Tsalev and Zaprianov, 1983; Evans et al., 1978).

The spectrophotometer used was a Perkin-Elmer, model 2380, with dual beam, background correction of deuterium (DBC). Hollow cathode tubes from Perkin-Elmer were used. Wavelengths for each element were: Pb 283.3 nm, Cu 324.8 nm, Cd 228.8 nm and Zn 213.9 nm. Recovery ranged from 82% to 90% for Pb, 88–92 for Cd, 95–100% for Cu and 92–101% for Zn. Method's detection limits were as follows: for Pb 0.1 mg/kg, Cd 0.05 mg/kg, Cu 0.2 mg/kg, Zn 0.2 mg/kg. Statistical analysis was performed using the method of one-way analysis of variance (ANOVA) (Armitage, 1971).

3. Results and Discussion

Copper (Cu) is the ingredient in a multitude of enzymes and plays an important role in many physiological functions of man and animals. Frequently though pathological conditions known as chalkoses with a characteristic high concentration of Cu, mainly in liver, kidneys and blood occur. These pathological states are caused by either an uptake of an excessive amount of Cu or by food containing normal amounts of Cu but low amounts of Mo or sulphur radicals (Doyle and Spaulding, 1978). In contrast, lack of copper causes disturbances with a characteristic example being swayback in sheep (Schuhmacher et al., 1991; Koh and Judson, 1986).

Mean concentration of Cu in liver in area A (107.8 µg/g) is more than double that in area B and significantly higher than area C (Table 1).

TABLE 1. Mean copper concentrations (µg/g, w/w[a]) in sheep liver and kidney

Area	A		B		C	
Tissue	N	Mean ± SE	N	Mean ± SE	N	Mean ± SE
Liver*	(26)	107.80 ± 8.47	(24)	52.40 ± 4.50	(20)	73.50 ± 8.50
Kidney	(26)	3.80 ± 0.18	(24)	3.70 ± 0.25	(20)	3.80 ± 0.09

[a]Wet weight

*P < 0. 01 statistically significant between areas

Furthermore, in area A, a percentage of 19.2 have over 150 µg/g of Cu in liver. This value borders the limit values for chalkosis, while in the other two areas there is no sample approaching these concentrations (Table 2). This is probably due to the spraying of fruit -tree cultures with copper containing compounds at the season where samples from these sheep were obtained (Koh and Judson, 1986).

In area B, as mentioned above, there were no samples having high Cu concentrations even though there was also spraying with copper containing chemicals. This must be attributed to the high sulphuric radical content in these soils which prevents copper from being absorbed.

In area C copper concentration ranged within normal levels. Copper concentrations in Canadian sheep and goats were 78.7 ppm in liver and 4.7 ppm in kidney (Salisbury and Chan, 1991). In Poland, copper concentrations were found to be 29 and 3.7 ppm in the liver and kidney, respectively (Falandysz, 1991).

Zinc (Zn) is a co-factor in many enzymes and plays a role in carbohydrate, lipid and protein metabolism. It is considered to be necessary for growth and development, in wound healing and in DNA synthesis (Peppa-Papasteriadou, 1986). It is comparatively non-toxic (Underwood, 1977) but pathological conditions often occur caused by lack of Zn as a result of consumption of food poor in Zn. Copper and phosphorus affect zinc absorption because they form non-absorbable complexes. A typical example of lack of zinc is parakeratosis in swine.

TABLE 2. Distribution (%) of copper concentrations (μg/g, w/w) in sheep liver

Area	A		B		C	
μg/g	N	%	N	%	N	%
10–100	13	50.0	23	95.8	16	80.0
100–150	8	30.8	1	4.2	4	20.0
>150	5	19.2	–	–	–	–
Total	**26**	–	**24**	–	**20**	–

If we assume that normal concentrations of Zn in the liver are 35–45 μg/g (Underwood, 1977) it is apparent that in all three areas zinc concentration in sheep tissues was low (Table 3). Investigators in Canada, Poland and USA report values of zinc concentration in liver and kidney of goat and sheep, ranged from 39 to 82.2 ppm and 23 to 147.2 ppm respectively (Falandysz, 1991; Salisbury and Chan, 1991; Khan et al., 1995a).

TABLE 3. Mean zinc concentrations (μg/g, w/w) in sheep liver and kidney

Area	A		B		C	
Tissue	N	Mean ± SE	N	Mean ± SE	N	Mean ± SE
Liver*	(26)	24.00 ± 2.17	(20)	26.50 ± 2.17	(20)	14.40 ± 0.47
Kidney*	(26)	18.50 ± 1.01	(24)	23.40 ± 1.86	(20)	10.30 ± 0.69

*P < 0.01 statistically significant between areas

Cadmium (Cd) is considered to be one of the most toxic heavy metals. In addition, cadmium is implicated in high blood pressure (Perry et al., 1979) prostate cancer, mutations, dysplasy and foetal (embryonic) death (Schuhmacher et al., 1991). Man and animals take up cadmium through air, water and mainly by food. Some foodstuff, in particular of animal origin, like liver and kidney, contain higher levels of Cd (Spierenburg et al., 1988).

Mean concentrations of Cd in liver were found to be 0.5 μg/g and in kidney below 1 μg/g (Table 4). These concentrations are considered to be at desirable levels (Spierenburg et al., 1988) except the ones in kidney from area B (1.23 μg/g) that marginally exceeded these levels but was well below the tolerable limits posed for human consumption (3 μg/g) (Spierenburg et al., 1988).

M. AL-ALAWI

TABLE 4. Mean cadmium concentrations (µg/g, w/w) in sheep liver and kidney

Area		A		B		C
Tissue	N	Mean ± SE	N	Mean ± SE	N	Mean ± SE
Liver*	(26)	0.17 ± 0.01	(24)	0.32 ± 0.03	(20)	0.18 ± 0.03
Kidney*	(26)	0.64 ± 0.09	(24)	1.23 ± 0.15	(20)	0.60 ± 0.06

*P < 0.01 statistically significant between areas

No liver samples contained cadmium more than 1 µg/g, while a small percentage had concentrations of 0.5–1 µg/g (Table 5). Only one kidney sample contained more than 3 µg/g, while the percentage at 1–3 µg/g (58.3%) only at the area B, was considered to be high (Table 6). A study of Khan et al. (1995b) reported concentrations of Cd in goat liver and kidney of 0.32 and 0.51 ppm, respectively, while Salisbury and Chan (1991) had found concentrations of 0.06 and 0.17 ppm, in tissues of goat and sheep, respectively.

TABLE 5. Distribution (%) of cadmium concentrations (µg/g, w/w) in sheep liver

Area		A		B		C
µg/g	N	%	N	%	N	%
<0.5	25	96.1	20	83.3	19	95.0
0.5–1	1	3.9	4	16.7	1	5.0
>1	–	–	–	–	–	–
Total	**26**		**24**		**20**	

TABLE 6. Distribution (%) of cadmium concentrations (µg/g, w/w) in sheep kidney

Area		A		B		C
µg/g	N	%	N	%	N	%
<1	23	88.5	9	37.5	18	90.0
1–3	3	11.5	14	58.3	2	10.0
>3	–	–	1	4.2	–	–
Total	**26**	–	**24**	–	**20**	–

The toxicity of lead (Pb) is attributed to the fact that it interferes with the normal function of a number of enzymes. Bipolar Pb forms strong bonds with enzymes bearing sulfhydryl groups thus inhibiting their action. Lead is toxic to blood and the nervous, urinary, gastric and genital systems. Furthermore, it is also implicated in causing carcinogenesis, mutagenesis and teratogenesis in experimental animals (Bryce-Smith and Stephens, 1983). The mean concentration of Pb in both tissues, in all areas, was at about the same level (0.71–0.96 µg/g) and this was considered to be a desirable level as it was less than 1 µg/g (Table 7). Finally, it was found that 16–20% of liver samples exceeded the desirable levels of 1 µg/g while the percentage in kidney was 15% in area C and was as high as 34.6% in area A (Tables 8 and 9).

TABLE 7. Mean lead concentrations (µg/g, w/w) in sheep liver and kidney

Area		A		B		C
Tissue	N	Mean ± SE	N	Mean ± SE	N	Mean ± SE
Liver*	(26)	0.80 ± 0.06	(24)	0.71 ± 0.07	(20)	0.82 ± 0.07
Kidney*	(26)	0.96 ± 0.07	(24)	0.72 ± 0.06	(20)	0.87 ± 0.04

*P > 0.01 not statistically significant between areas

TABLE 8. Distribution (%) of lead concentrations (μg/g, w/w) in sheep liver

Area	A		B		C	
μg/g	N	%	N	%	N	%
<0.5	4	15.4	8	33.3	5	25.0
0.5–1	17	65.4	12	50.0	11	55.0
>1	5	19.2	4	16.7	4	20.0
Total	**26**	–	**24**	–	**20**	–

TABLE 9. Distribution (%) of lead concentrations (μg/g, w/w) in sheep kidney

Area	A		B		C	
μg/g	N	%	N	%	N	%
<0.5	3	11.5	8	33.3	1	5.0
0.5–1	14	53.9	12	50.0	16	80.0
>1	9	34.6	4	16.7	3	15.0
Total	**26**	–	**24**	–	**20**	–

4. Conclusion

Sheep tissue concentration of Cu in area A was high, while in the other two areas it was at normal levels. This high concentration must be attributed mainly to the spraying with copper compounds at fruit-tree cultures. It is therefore wise to avoid grazing at these areas during the season of spraying. Zinc concentration in sheep liver and kidney was considered to be at low levels. Therefore, addition of Zn in feed is recommended. Cadmium concentration was at desirable levels. Finally, lead concentration was at about the same level in all three areas and these values considered desirable.

References

Armitage, P., 1971, Statistical methods in medical research. Blackwell, Oxford

Bryce-Smith, D., Stephens, R., 1983, Sources and effects of environmental lead. In: Trace elements in health, J. Rose (Ed). A review of current issues. Butterworths, London, pp. 112–114

Doyle, J.J., Spaulding, J.E., 1978, Toxic and essential trace elements in meat. A review, *J. Anim. Sci.*, 47: 398–419

Ellinder, G., Kjelstrom, T., 1977, Cadmium concentration in samples of human kidney cortex from 19th Century, *Ambio*, 6: 270

Evans, W.H., Read, J.J., Lucas, B.E., 1978, Evaluation of a method for the determination of total cadmium lead and nickel in foodstuffs using measurement by flame atomic absorption spectrophotometry, *Analyst*, 103: 580

Falandysz, J., 1991, Manganese, copper, zinc, iron, cadmium, mercury and lead in muscle, meat, liver and kidneys of poultry, rabbit and sheep slaughtered in the Northern part of Poland, *Food Add. Contam.*, 8: 71–83

Jacobson, K., Turner, J., 1980, The Interaction of cadmium and certain other ions with proteins and nucleic acids, *Toxicolology*, 16: 1–37

Khan, T.A., Diffay, C.B., Forester, M.D., Thompson, J.S., Mielke, W.H., 1995a, Trace element concentrations in tissues of goats from Alabama, *Vet. Human Toxicol.*, 37: 327–329

Khan, T.A., Diffay, C.B., Datiri, C.B., Forester, M.D., Thompson, J.S., Mielke, W.H., 1995b, Heavy metals in livers and kidneys of goats in Alabama, *Bull. Environ. Contam. Toxicol.*, 55: 568–573

Koh, T.S., Judson, G.J., 1986, Trace elements in sheep grazing near a lead-zinc smelting complex at Port Pirie, South Australia, *Bull. Environ. Contam. Toxicol.*, 37: 87–95

Peppa-Papasteriadou, M., 1986, The trace element concentrations of Zn, Cu, Mn and Se in children and adolescents. Ph.D. Thesis. Aristotelian University of Thessaloniki, Medicine Faculty, Thessaloniki Greece, pp. 13–23

Perry, H.M., Erlanger, M., Perry, F., 1979, Increase of the systolic pressure of rats chronically fed cadmium, *Environ. Health Perspect.*, 28: 261

Salisbury, C.D.C., Chan, W., 1991, Multielement concentrations in liver and kidney tissues from five species of Canadian slaughtered animals, *J. Assoc. Off. Anal. Chem.*, 74: 587–591

Schuhmacher, M., Bosque, A.M., Domingo, L.J., Corbella, J., 1991, Dietary intake of lead and cadmium from foods in Tarragone province, *Spain Bull. Environ. Contam. Toxicol.*, 46: 320–328

Spierenburg, J., De Graaf, J.G., Baars, J.A., Brus, D.H.J., Tielen, M.J.M., Arts, B.J., 1988, Cadmium, Zinc, Lead and Copper in livers and kidneys of cattle in the Neighbourhood of Zinc refineries, *Environ. Monit. Assess.*, 11: 107–114

Tsalev, L.D., Zaprianov, Z.K., 1983, Atomic absorption spectrometry in occupational and environmental health practice, CRC, Boca Raton, FL

Underwood, J.E., 1977, Trace elements in human and animal nutrition, 4th ed. Academic, London

TRACE ELEMENTS AND HEAVY METALS STATUS IN ARABIAN CAMEL

BERNARD FAYE
CIRAD-ES, Campus International de Baillarguet, Montpellier, France

RABIHA SEBOUSSI, MOSTAFA ASKAR
UAE university, Al-Ain, United Arab Emirates

Abstract: In the desert, camel rearing is an important cultural fact. In the present paper, 240 Arabian camels from Emirates were sampled for the determination of trace elements and different heavy metals. The following elements were tested: copper, zinc, iron, aluminium, arsenic, boron, barium, cobalt, chromium, cadmium, manganese, molybdenum, nickel, selenium, strontium and lead. The variation factors included age, sex and physiological status. On the average, the mineral contents were 190.3 µg/100 ml (iron), 60.1 µg/100 ml (copper), 44.0 µg/100 ml (strontium), 22.5 µg/100 ml (arsenic), 20.0 µg/100 ml (zinc), 19.7 µg/100 ml (selenium), 19.3 µg/100 ml (boron) and 14.6 µg/100 ml (barium). Other minerals like aluminium (3.7 µg/100 ml), molybdenum (2.9 µg/100 ml), chromium (2.0 µg/100 ml), nickel (1.8 µg/100 ml), lead (1.5 µg/100 ml), manganese (0.16 µg/100 ml), cobalt (0.08 µg/100 ml) and cadmium (0.07 µg/100 ml) were in very small concentration. Age, sex and physiological effects were assessed for some parameters. According to the lack of references in camel species, it is difficult to link those results to polluting context. But those data could contribute to understand the heavy metal status in camel confronted to pollution.

Keywords: Heavy metal, trace element, camel blood, racing camel

1. Introduction

The main trace elements in camel serum (copper, zinc, iron) were commonly determined in countries where camels play an important role in the livestock economy. Some reviews are available in the literature (Faye and Bengoumi, 1994, 2000; Abu Damir, 1998). Normal ranges and deficiency statuses are described in numerous cases. However, the data on other trace elements and heavy metals are scarce. The importance of those other trace elements and their potential toxicity were widely described in other species as small ruminants or cattle. In camel, the references are scarce and even non available for some elements. In Arab Emirates, the racing camel has a central cultural place. Racing activities have strong effect on the metabolism and physiology of the camel (Rose et al., 1994). Those animals have specific mineral requirements under effort. The determination of a wide type of trace elements could be beneficial for a better understanding of the specific physiology of sport animal in desert conditions.

In the present paper, copper, zinc, iron which are classically determined in camel blood are enriched by the determination of other minerals as aluminium, arsenic, boron, barium, cobalt, chromium, cadmium, manganese, molybdenum, nickel, selenium, strontium and lead. Some of these elements are biologically essential and some others are potentially toxic and linked mainly to industrial pollution.

2. Materials and Methods

2.1. ANIMALS

The animals were provided by Al-Ochouche farm at Al-Ain Emirates including 3,000 dromedary camels (*Camelus dromedarius*) in extensive management of three breeds adapted to race: local, Sudanese and crossbred. For the present study, 240 animals between 2 and 10 years old were randomly selected. Before blood collecting, a general examination of all the selected camels was achieved and only healthy animals were retained. To discard the animals with trypanosome, common disease in Arab Emirates, a diagnosis was performed by three different tests: mercury chlorate test, agglutination test and blood examination test according to Woo method (Woo, 1971). Each positive animal at one of the test was discarded. A faecal examination for internal parasites diagnosis was achieved according floating method (Soulsby, 1982).

Finally, the analyses were achieved in 235 animals. The camel samples were shared into two groups according the gender. The sample included 83 males and 152 females, a part of the she-camels was pregnant (n = 68), 55 non pregnant and 29 lactating. Each dromedary camel was identified by a number tied to the neck. The males and females animals were distributed into three age classes, respectively 3–4 years, 5–7 years and 8 years and more. The males were mainly less than 4 years old (n = 51) or mature breeding animals more than 8 years old (n = 32). Females less than 4 years old were 49. The number of females in other classes was respectively 97 (class 5–7 years) and 6 (8 and more). On the whole population, 156 were local breeds, 6 Sudanese only and 73 crossbreed. So, the experimental design allowed the assessment of gender, age, breed and physiological stage effect.

Animals were fed with a basal diet including green alfalfa, commercial concentrate, dates and a mixture of three types of lentils. They were watered *ad libitum*. No specific mineral supplementation was given for the experimental time except Selenium and vitamin E regularly introduced in the commercial supplement, especially for pregnant camels.

2.2. BLOOD SAMPLING

After disinfection of the skin at the upper part of the neck with iodine alcohol, blood was collected at jugular vein with sterile syringe in two 10 ml tubes without EDTA. Those samples were carried later to the laboratory for analyses performed immediately after collecting.

The collected samples were centrifuged at 4,300 t/min for 5 min. The main trace elements (Zn, Cu, Fe) were determined after separation of serum by atomic absorption spectrophotometer according to the classical method of Bellanger and Lamand (1975). The kit used came from company *Dade Boehring USA*. Those analyses were achieved at the veterinary lab of agricultural department at Al-Ain.

The other mineral analysis (Al, As, B, Ba, Cd, Co, Cr, Mn, Mo, Ni, Pb, Se, Sr) were achieved on serum stored at 4°C before analysis according to the Brown and Watkinson method (1977). The samples were digested for destroying proteins and amino acids in order to separate the minerals linked to proteins. This first step was performed at the private department of H.H. Sheikh Zayed Bin Sultan Al-Nahyan, in the scientific centre of racing camel, Al-Ain. The concept was to mix in the 6 tubes of the rotator in a microwave digestor, 2 ml serum, 6 ml hydrogen peroxide (H_2O_2) at 30%, then 1 ml nitric acid (HNO_3) at 60%. The tubes was placed in the rotator by increasing order from 1 to 6 and well tightened, then introduced in the apparatus. After serum digestion, the sample was poured in sterile tube, then exported to Al-Salamate lab analysis –Al Ain for determining the minerals with an ICP (Induced coupled plasma *Varian Vista MPX-CCD*).

2.3. STATISTICAL ANALYSIS

The differences between groups (age classes, sex groups, physiological status groups, breeds) were tested by variance analysis according to the procedure General Linear Models (GLM) with R software©. If the probability (p) was below 0.05, the differences between groups were considered as significant. The correlation of Pearson between the analyzed elements was calculated. The interactions between variation factors were taken in account in statistical models.

3. Results

In order to facilitate comparisons, all the results were expressed in $\mu g/100$ ml. On average, the main trace element in camel serum were iron (190.3 $\mu g/100$ ml) followed by copper (60.1 $\mu g/100$ ml), strontium (44.0 $\mu g/100$ ml), arsenic (22.5 $\mu g/100$ ml), zinc (20.0 $\mu g/100$ ml), selenium (19.7 $\mu g/100$ ml), boron (19.3 $\mu g/100$ ml) and barium (14.6 $\mu g/100$ ml). Other minerals like aluminium (3.7 $\mu g/100$ ml), molybdenum (2.9 $\mu g/100$ ml), chromium (2.0 $\mu g/100$ ml), nickel (1.8 $\mu g/100$ ml), lead (1.5 $\mu g/100$ ml), manganese (0.16 $\mu g/100$ ml), cobalt (0.08 $\mu g/100$ ml) and cadmium (0.07 $\mu g/100$ ml) were in very small concentration.

3.1. AGE EFFECT

There was no observed age effect on Al, Ba, Cd, Co, Cr, Cu, Mn, Mo, Ni, Pb and Sr (Table 1). When an age effect was reported (Table 1), the young adult animals (5–7 years old) had a significant higher concentration of arsenic (28.3 $\mu g/100$ ml), boron (35.8 $\mu g/100$ ml) and selenium (28.1 $\mu g/100$ ml). The oldest animals more than 8 years old had highest iron values (283 $\mu g/100$ ml). However, the differences between age groups appeared slight. Concerning zinc, a significant lower value was reported in group 2 (5–7 years old).

TABLE 1. Mean values of trace elements and heavy metals in camel serum according to age groups (in $\mu g/100$ ml)

Minerals	Age group		
	3–4 Years	5–7 Years	>7 Years
Al	2.1	6.6	0.7
As	18.7	28.3**	17.9
B	7.1	35.8**	9.3
Ba	15.4	14.5	12.7
Cd	0.06	0.09	0.16
Co	0.07	0.02	0.28
Cr	2.5	1.6	1.8
Cu	57.7	62.9	59.0
Fe	174.1	171.5	283.8**
Mn	0.2	0.07	0.29
Mo	3.0	3.6	1.2
Ni	1.8	1.7	1.9
Pb	1.6	1.2	1.8
Se	14.1	28.1**	12.4
Sr	44.0	44.6	42.3
Zn	25.6	12.1**	25.1

*P < 0.05; **P < 0.01

3.2. SEX EFFECT

Some elements (As, Ba, Cd, Co, Cr, Mn, Mo, Pb) had similar values between sex. Females had significant higher values for boron (23.3 µg/100 ml), copper (61.9), selenium (22.9) and strontium (47.6). Aluminium was also slightly higher in female (4.6 µg/100 ml). Iron value was significantly higher in male (213.1 µg/100 ml), as well as zinc (24.3 µg/100 ml). Nickel was slightly higher in male also (2.0 *vs* 1.7) (Table 2).

TABLE 2. Mean values of trace elements and heavy metals in camel serum according to sex groups (in µg/100 ml)

Minerals	Male	Female
Al	2.1	4.6*
As	22.3	22.6
B	11.9	23.3**
Ba	14.7	14.6
Cd	0.17	0.02
Co	0.16	0.04
Cr	2.6	1.6
Cu	56.7	61.9**
Fe	213.1**	177.8
Mn	0.24	0.12
Mo	2.2	3.3
Ni	2.0*	1.7
Pb	1.7	1.3
Se	13.6	22.9**
Sr	37.4	47.6**
Zn	24.3**	17.6

*P < 0.05; **P <0.01

3.3. PHYSIOLOGICAL STATUS EFFECT

As for former variation factors, no effect of the physiological status was observed for Cd, Co, Cr, and Mo. Elsewhere, Ba, Cu, Mn, and Ni did not change according to the status of the female (non pregnant, pregnant or milking). In non pregnant animals, strontium (52.9 µg/100 ml), iron (189.1 µg/100 ml) and zinc (27.3) were in higher concentrations in plasma while selenium was significantly lower (13.8). In pregnant animals, values of copper were slightly higher but not significantly (64.7 µg/100 ml). In milking animals, significant higher values of arsenic (32.4 µg/100 ml), boron (51.7 µg/100 ml) and aluminium (10.2 µg/100 ml) were observed (Table 3).

3.4. BREED EFFECT

Few elements changed according to the breed (Table 4). The observed differences were slightly significant. Boron was slightly lower in crossbreed group (11.7 µg/100 ml *vs* 22.7 in other groups). Arsenic was also lower in crossbreed group (19.8 µg/100 ml *vs* 23.8 and 21.5 in the other breeds).

TABLE 3. Mean values of trace elements and heavy metals in camel serum according to physiological status group (in µg/100 ml)

Minerals	Physiological status group		
	No-preg	Pregnant	Milking
Al	1.0	5.1	10.2**
As	12.7	26.5	32.4**
B	1.3	29.1	51.7**
Ba	14.7	15.7	11.7
Cd	0.02	0.03	0.00
Co	0.06	0.03	0.00
Cr	1.8	1.6	1.5
Cu	60.2	64.7	58.9
Fe	189.1*	174.7	163.9
Mn	0.19	0.10	0.00
Mo	2.9	2.5	6.2
Ni	1.7	1.7	1.9
Pb	1.6*	1.7	1.7
Se	13.8**	28.1	28.2
Sr	52.9**	45.9	41.5
Zn	27.3**	13.5	8.9

$*P < 0.0; **P < 0.01$

TABLE 4. Mean values of trace elements and heavy metals in camel serum according to breed groups (in µg/100 ml)

Minerals	Breed group		
	Local	Crossbred	Sudanese
Al	4.3	2.0	9.7
As	23.8*	19.8	21.5
B	22.7	11.7*	22.7
Ba	14.6	14.8	13.9
Cd	0.06	0.02	0.24
Co	0.07	0.02	0.28
Cr	2.1	1.7	1.9
Cu	61.0	58.3	58.4
Fe	198.3	176.3	153.4
Mn	0.12	0.21	0.54
Mo	2.5	3.7	5.8
Ni	1.8	1.7	1.9
Pb	1.4	1.4	3.2
Se	19.6	19.5	20.9
Sr	45.1	40.6	56.4
Zn	19.0	22.0	21.0

$*P < 0.05; **P < 0.01$

3.5. INTERACTIONS

The interaction age*sex was observed for boron and selenium only ($p < 0.01$), essentially because no males belonged to age group 2 where highest values of B and Se were observed.

4. Discussion

Except for copper, zinc, iron and in a less extent selenium, the references concerning trace element concentrations in camel blood, serum or plasma are quite marginal. The discussion will concern first the main trace elements as copper, zinc, manganese, iron and selenium, secondly the rarely analyzed trace elements (aluminium, boron, barium, chrome, cobalt, molybdenum, nickel and strontium) and last the toxic minerals (arsenic, cadmium, lead).

4.1. THE MAIN TRACE ELEMENTS

Serum or plasma **copper** is a good reflect of copper intake. In ruminants, normal copper concentrations are between 70 and 120 µg/100 ml (i.e. 12 and 19 µmol/l). Most of the reported values in camel were inside those thresholds (Faye and Bengoumi, 1994). With a mean value closed to 60 µg/100 ml, the copper status of the camel in our study was at the deficiency limit. In the literature, no significant variation due to sex was reported (Abdalla et al., 1988; Bengoumi et al., 1995), but the change along the gestation was observed (Liu et al., 1994) with a decrease of copper concentration at the end of pregnancy, contrary to our results. The results concerning age effect are contradictory: no significant difference (Faye and Mulato, 1991; Bengoumi et al., 1995), higher value on camels more than 5 years old (Marx and Abdi, 1983) as for the present study.

As for copper, **zinc** concentration in plasma or serum for most of the ruminants is between 70 and 120 µg/100 ml. The present results confirm a different pattern for camel. Indeed, normal values in camel were around 30–50 µg/100 ml (Faye et al., 1992). In the present study, the zinc values was below these limits in most of the cases and varied between 0.2 and 115 µg/100 ml. Low values were already reported in Emirates (Abdalla et al., 1988) and zinc deficiency was commonly suspected in the camel stock from this part of the world.

The age and sex variations of plasma or serum zinc were rarely reported. Young camels below 2 years showed generally lower values (Faye et al., 1995). The highest values observed by some authors on non weaning camel calves were due to the milk feeding which provided sufficient zinc in the diet. A decreasing of zinc concentration was observed at the end of gestation (Faye and Mulato, 1991) as we reported in the present study. This decreasing could be linked to an active transfer of plasma zinc to the foetus. In Bactrian camel, similar trends were observed in pregnant animals (Liu et al., 1994). In our sample, the values in milking camel were very low, but no clinical symptoms of zinc deficiency were observed. These very low values, probably due to zinc transfer into the milk, could explain the sex difference reported in our sample, the lowest values being reported in milking animal. A recent study has shown an active transfer of zinc into camel milk (Cattaneo et al., 2005).

Manganese can be a limiting factor of the mineral diet in ruminants and deficiencies can be locally present according to the low concentration in some grasses from southern countries (Faye et al., 1986). Usually the quantity of manganese in blood was very low and can be detected only recently with high accuracy thanks to sophisticated apparatus as ICP. The mean values reported in the literature vary from 8.4 µg/100 ml in Morocco (Bengoumi et al., 1994) to 30 µg/100 ml in Egypt (Eltohamy et al., 1986). In ruminants, the values of blood manganese concentration are generally below 10 µg/100 ml (Lamand, 1987). No variations factors were reported in the literature. In our sample, the values were effectively very low with a wide range (1.64 ± 6.69 µg/ml) and very slight difference was observed according to physiological status. Probably, the transfer to foetus, then to milk could explain the low plasma values in pregnant and milking female.

Iron is a common element of the nature, especially in tropical conditions. It is the most important element in the blood which contributes to haemoglobin composition. The references on plasma, serum or whole blood iron in camel are common. In serum, the values varied on average from 98 µg/100 ml (Tartour, 1970) to 186 µg/100 ml (Moty et al., 1968). In Emirates, Abdalla et al. (1988) reported mean values at 113 µg/100 ml. Our results are in the upper limit of those published data.

As for other main trace elements, iron decreased in pregnant and milking camel (Eltohamy et al., 1986). However, this change was not significant in our results. The iron concentration was generally higher in adult camels (Marx and Abdi, 1983; Shekhawat et al., 1987; Ghosal and Shekhawat, 1992) as in our sample. Concerning sex effect, the results in the literature were quite contradictory (Faye and Bengoumi, 1994). Our results mentioning a higher mean value in males were only in accordance with those of Hussein et al. (1997).

The **selenium** deficiency was described in young camel by several authors (Finlayson et al., 1971; Hamliri et al., 1990; Musa and Tageldin, 1994). In whole blood, Hamliri et al. (1990) reported values between 10.9 and 11.8 µg/100 ml in Morocco. No age or sex effect was observed. Similar values involving Bactrian camels were published in China (Liu et al., 1994) with 9.7–11.4 µg/100 ml. Higher values (28.1 µg/100 ml) were observed in camel plasma from Oman (unpublished data).

In a trial including a supplementation period (Bengoumi et al., 1998), the mean plasma selenium concentration in camel was 2.1 µg/100 ml (before supplementation), 12.9 (during supplementation) and 8.3 (after supplementation period). The maximum mean value was observed the day before the end of supplementation period: 20.1 µg/100 ml. According to Liu et al. (1994), selenium concentration did not change according to the physiological status. In Emirates, selenium supplementation in pregnant camels was common. This practice explained the highest values reported in pregnant and milking females, and in age group 2 (due to sex*age interaction).

4.2. RARE TRACE ELEMENTS AND HEAVY METALS

The **cobalt** concentration was generally very low in the plasma (Lamand, 1987) and did not reflect the cobalt status contrary to cyanocobalamin (Vitamin B12). There was no data on cobalt in dromedary camel blood. Some results were available in Bactrian camel from Mongolia and China. According to Burenbayar (1989), cobalt concentration in blood varied from de 3.4 to 13.2 µg/100 ml according to the season and mineral supplementation but the analysis method was not given. By atomic absorption spectrometry, Liu et al., (1994) reported blood cobalt concentration at 39 µg/100 ml in non-pregnant camel, 56 µg/100 ml in pregnant female and 53 µg/100 ml after parturition, without significant difference. These values are quite higher than our results (0.08 µg/100 ml in average). The analyzing method could be debatable for the previous studies.

Molybdenum is generally in competition with copper. Excess of molybdenum associated with sulphur is known to decrease copper digestibility in ruminants. Some cases of molybdenosis were described in camels grazing bush with *Salvadora persica* as predominant plant (Faye and Mulato, 1991). In Bactrian camel blood, molybdenum concentration was between 19 and 23 µg/100 ml according to Liu et al. (1994) and 0.43– 0.53 µg/100 ml only for Ma (1995). Our values are intermediate between those publicshed results. No pregnancy effect was observed in Bactrian camel (Liu et al., 1994).

No **nickel** plasma values were reported in camel, in spite of the recent interest for nickel in ruminants. In Mongolia, a "roll disease" linked to nickel intoxication was described in Bactrian camel (Tao et al., 1995).

Lead is a toxic element which presents an interest as indicator of environmental pollution. No case of lead intoxication was described in camel. In Saudi Arabia, Elamin and Wilcox (1992) reported considerable lead concentration in camel milk (180 µg/g DM). In cow blood, lead concentration was generally between 0.6 and 4.8 µg/100 ml (Jeffrey et al., 2003).

No data was available for the other elements in camel. **Aluminium, chromium** and **strontium** were analyzed in muscle and hump fat, but only traces were observed. In dairy cow, blood concentration of chromium in non supplemented animals was between 0.33 and 0.42 µg/100 ml according to Pechova et al. (2002). Generally in cow, **cadmium** was below 0.1 µg/100 ml.

5. Conclusion

Except for the main trace elements, very few data or even no data were available to compare those results to the literature concerning camel. The camel seems to be less efficient than other ruminants as the goat to detoxify its organism (Al-Qarawi and Ali, 2003), so the sensitivity of camel to some toxic elements could be more important. It is expected that other determinations of heavy metals and toxic elements in blood and other biological fluids will be achieved to enlighten the standard values in this species.

Acknowledgements

We are grateful to the institutions and their directors which gave us all facilities to achieve the analysis: (i) Municipality and Agriculture Department – His Excellency Jumaa Saeed Hareb Al Amimi, (ii) Scientific Center for Racing Camel. Dr. Jawad Al Massri, (iii) Al Salamat Laboratory- Eng .Omar.

References

Abdalla, O.M., Wasfi, I.A., Gadir, F.A., 1988, The Arabian race normal parameters. I. Haemogram enzymes and minerals, *Comp. Biochem. Physiol.*, 90A: 2337–239

Abu Damir, H., 1998, Mineral deficiencies, tonicities and imbalances in the camel (Camelus dromedaries): a review, *Vet. Bull.*, 68: 1103–1119

Al-Qarawi, A.A., Ali, B.H., 2003, Variations in the normal activity of esterases in plasma and liver of camels, cattle, sheep and goats, *J. Vet. Med.*, A50: 201–203

Bellanger, J., Lamand, M., 1975, Méthodes de dosage du cuivre et du zinc plasmatique, *Bull. Tech. CRZV. Theix Inra*, 20: 53–54

Bengoumi, M., Faye, B., Elkasmi, K., and Tressol, J.C., 1995, Facteurs de variation des indicateurs plasmatiques du statut nutritionnel en oligo-element chez le dromadaire au Maroc. II. Effect d une complementation minerale, *Rev. Elev. Med. Vet. Pays. Trop.*, 48: 277–280

Bengoumi, M., Essamadi, A.K., Tressol, J.C., Charconac, J.P., Faye, B., 1998, Comparative effect of selenium concentration and erythrocyte glutathione peroxidase activity in cattle and camels, *Anim. Sci.*, 67: 461–466

Brown, M.W., Watkinson, J.H., 1977, An automated fluorometric method for the determination of nanograms quantities of selenium, *Anal. Chem. Acta*, 89: 29–35

Burenbayar, R., 1989, Supply of trace-elements for female camels in Mongolia, VIth Int. Trace-element symp., Leipzig, DDR, 2

Cattaneo, D., Dell'Orto, V., Fava, M., Savoini, G., 2005, Influence of feeding on camel milk components. Proc. of Intern. Workshop, Desertification combat and food safety: the added value of camel producers". Ashkabad (Turkménistan), 19–22 April 2004. In "Vol. 362 NATO Sciences Series, Life and Behavioural Sciences", B. Faye and P. Esenov (Eds). IOS press, Amsterdam, The Netherlands, 181–186

Elamin, F.M., Wilcox, C.J., 1992, Milk composition of Marjheim camels, *J. Dairy Sci.*, 75: 3155–3157

Eltohamy, M.M., Salama, A., Youssef, A.E.A., 1986, Blood constituents in relation to the reproductive state in she-camel (*Camelus dromedarius*), *Beitrage zür Trop. Landwirtschaft und Vet. Med.*, 24: 425–430

Faye, B., Grillet, C., Tessema, A., 1986, Teneur en oligo-éléments dans les fourrages et le plasma des ruminants domestiques en Ethiopie, *Rev. Elev. Méd. Vét. Pays Trop.*, 39: 227–237

Faye, B., Mulato, C., 1991, Facteurs de variation des paramètres protéo-énergétiques, enzymatiques et minéraux dans le plasma chez le dromadaire de Djibouti, *Rev. Elev. Méd. Vét. Pays Trop.*, 44: 325–333

Faye, B., Bengoumi, M., 1994, Trace-element status in camel. A review, *Biol. Trace Elem. Res.*, 41: 1–11

Faye, B., Saint-Martin, G., Cherrier, R., Ali Ruffa, M., 1992, The influence of high dietary protein, energy and mineral intake on deficient young camel: II. Changes in mineral status, *Comp. Bioch. Physiol.*, 102A: 417–424

Faye, B., Bengoumi, M., 2000, Le dromadaire face à a sous-nutrition minérale: un aspect méconnu de son adaptabilité aux conditions désertiques, *Sécheresse*, 11(3): 155–161

Finlayson, R., Keymer, I.F., Manton, J.A., 1971, Calcific cardiomyopathy in young camels (*Camelus* spp.), *J. Comp. Path.*, 81: 71–77

Ghosal, A.K., Shekhawat, V.S., 1992, Observations on serum trace elements levels (zinc, copper and iron) in camel (*Camelus dromedarius*) in the arid tracts of Thar Desert in India, *Rev. Elev. Méd. Vét. Pays Trop.*, 45: 43–48

Hamliri, A., Khallaayoune, K., Johnson, D.W., Kessabi, M., 1990, The relationship between the concentration of selenium in the blood and the activity of glutathione peroxidase in the erythrocytes of the dromedary camel (*Camelus dromedarius*), *Vet. Res. Comm.*, 14: 27–30

Hussein, M.F., Al Mufanej, S.I., Mogarve, H.H., Le Nabi, A.R.G., Sanad, H.H., 1997, Serum iron and iron binding capacity in the arabian camel (*Camelus dromedarius*), *J. Appl. Anim. Res.*, 11: 201–205

Jeffrey, S.L., Whitaker, S.M., Borger, D.C., Willett, L.B., 2003, Pharmacokinetics of lead in cattle: transfer to calf, *Anim. Sci. and Rev.* Special circular, 156

Lamand, M., 1987, Place du laboratoire dans le diagnostic des carences en oligo-éléments, *Rec. Med. Vet.*, 163: 1071–1082

Liu, Z.P., Ma, Z., Zhang, Y.J., 1994, Studies on the relationship between sway disease of Bactrian camels and copper status in Gansu Province, *Vet. Res. Comm.*, 18: 251–260

Ma, Z., 1995, Studies on sway disease of chinese bactrian camels. Epidemiological and aetiological aspects, Report of International Foun dation for Science Project, Stockholm, Sweden, 17

Marx, W., Abdi, H.N., 1983, Serum levels of trace elements and minerals in dromedaries (Camelus dromedarius) in South Somalia, *Anim. Res. Dev.*, 17: 83–90

Moty, I.A., Mulla, A., Zaafe, S.A., 1968, Copper, iron and zinc in the serum of egyptian farm animals. *Sudan Agric. J.*, 3: 146–151

Musa, B.E., Tageldin, M.H., 1994, Swelling disease of dromedary camel, *Vet. Rec.*, 135: 416

Pechova, A., Podhorsky, A., Lokajova, E., Pavlata, L., Illek, J., 2002, Metabolic effects of chromium supplementation in dairy cows in the peripartal period, *Acta Vet. Brno*, 71: 9–18

Rose, R.J., Evans, D., Henkel, P., Knight, P.K., Cluer, D., Saltin, B., 1994, Metabolic responses to prolonged exercise in the racing camel. In: The racing camel. Physiology, metabolic functions and adaptations (chap. VII), B. Saltin and R.J. Rose (Eds). Suppl. *Acta Physiol. Scand.*, 150: 49–60

Shekhawat, V.S., Bathia, J.S., Ghosal, A.K., 1987, Serum iron and total iron capacity in camel. Indian, *J. Anim. Sci.*, 57: 168–169

Soulsby, E.J.L., 1982, Helminths, Arthropods and protozoa of domesticated animals, 7th ed. Bailliere Tnindal, London

Tao, G., Geng, C., Tao, Y., Jiao, H., Fan, Y., 1995, "Roll disease" of camels in Mongolia, *Acta Vet. Zootechn. Sinica*, 26: 541–544

Tartour, G., Idris, O.F., 1970, Studies of copper and iron metabolism in the camel fetus. *Acta Vet. (Brno)*, 39: 397–403

Woo, P.T.K., 1971, The hematocrit centrifuge technique for the detection of low virulent strains of trypanosomes of Trypanosome congolense subgroup, *Acta. Trop.*, XXVII 304–308

PLANT, WATER AND MILK POLLUTION
IN KAZAKHSTAN

EMILIE DIACONO, BERNARD FAYE
Département environnement et société, CIRAD, Montpellier, France

ALIYA MELDEBEKOVA, GAUKHAR KONUSPAYEVA
Kazakh State University Al Farabi, Almaty, Kazakhstan

Abstract: Since its independence in 1991, Kazakhstan is in a state of "ecological crises", due to the specific place for nuclear test by soviet government for long time, and to the development of irrigation for field cottons linked with decreasing Aral Sea level. In addition the manufacturing of metals and the minerals had some impact on environmental contamination.

In the South of Kazakhstan, eight farms were sampled close to probable pollution sources. Samples of camel milk, fodder and water were collected in each farm and analyzed for copper, iron, manganese, zinc, arsenic and lead. The mean content in fodder of Cu, Fe, Mn, Zn, As and Pb was 10.40 ± 2.93, 793.69 ± 630.48, 62.38 ± 20.67, 32.95 ± 27.15, 1.03 ± 0.49 and 4.28 ± 9.60 ppm respectively. In camel milk mean content of these heavy metals was respectively of 0.07 ± 0.04, 1.48 ± 0.53, 0.08 ± 0.03, 5.16 ± 2.17, <0.1, and 0.025 ± 0.02 ppm respectively. No heavy metals were detected in samples of water with the analytical methods used. .

The relationships between heavy metals in water, forages and milk were not clear. Some information's are lacking. We need to extend sampling at more areas where camels, cows, goat and sheep farms are closed to pollution areas, and analyzed other heavy metals suspected in pollution process.

Keywords: Camel milk, trace elements, heavy metals, fodder grass, pollution

1. Introduction

Kazakhstan has to face to important ecological problems due to nuclear tests, used of pesticides, polymetal industries, spatial base and traffic road increasing. In addition, in some areas there is overgrazing, decrease of water sources (Aral Sea for example) and their contamination by human activities (wastes of industries, pesticides residues...).

In literature, some references on heavy metals content in cow and breast milk are available. But concerning sheep and goat milk there are few references, and no one on camel milk.

According to the literature, there are some factors which influence the concentration of heavy metals in milk. These factors are in generally the levels and kind of human activities in the area of study. In that case, traffic road intensity plays a role on lead content in cow milk. In fact, the lead content in milk was positively correlated to the traffic density, (from 0.36 ppm on average for a traffic density of 10 vehicles per day to 7.20 ppm on average for a traffic density of 15,000 vehicles per day) (Bhatia and Choudhri, 1996).

Some studies showed significant difference on lead, arsenic, zinc, copper, iron concentration in milk in function of human activities near the sampling areas. Thus, lead concentration in cow milk was on average of 0.00132 ppm in rural area (Licata et al., 2003) and of 0.25 ppm in industrial area (Swarup et al., 2005) and 0.032 ppm close a road (Simsek et al., 2000). In Germany and Holland, higher value allowed is 0.05 ppm of Pb in milk, 0.02 ppm in Turkey and 0.1 ppm in Kazakhstan.

In industrial area, lead concentration of cow milk varied from 0.049 ppm (Simsek et al., 2000) to 0.067 ppm (Dey and Swarup, 1996), with higher value of 0.844 ppm on average, obtained near zinc and lead smelter (Swarup et al., 2005).

Concerning others heavy metals as arsenic, cadmium, zinc, copper, chromium and so on few references were available. Arsenic concentration in milk varies in function of sampling place. Thus, cows in rural areas produced milk containing 0.0002 ppm arsenic on average, in industrial area and traffic road area, milk samples contained 0.04 and 0.05 ppm on average (Simsek et al., 2000). The higher reported values were from 0.12 ppm in industrial area (Simsek et al., 2000) to 0.684 ppm (Licata et al., 2004).

For cadmium, 0.0345 ppm in cow milk (Fursov et al., 1985) was the highest value reported in literature. Tripathi et al. (1999) reported $0.07.10^{-3}$ ppm and Vidovic et al. (2005) found $0.0071.10^{-3}$ ppm. One reference on camel milk, near Aral sea, reported 0.1 ppm Cd in whole fresh and fermented milk (Saitmuratova et al., 2001).

Concerning zinc, it seems to be every time in the normal range in milk even if animals are rearing near industrial or traffic areas. Although, milk from industrial contained significantly more zinc than in traffic and rural area, 5.01 ppm, 4.49 ppm and 3.77 ppm respectively (Simsek et al., 2000).

For copper, observations were similar: the proximity of industrial area increased significantly the copper concentration in cow milk, as well as traffic area, cow milk from rural area contained less copper (0.96 ppm, 0.58 ppm and 0.39 ppm respectively) (Simsek et al., 2000).

For iron, human activities as industrial plants, traffic road, increased significantly the content in cow milk towards rural area (4.27 ppm, 1.78 ppm, 1.01 ppm respectively) (Simsek et al., 2000)

These results show that the type and the level of human activities in cattle rearing area could increase content of heavy metals and trace element in milk.

Studies on mercury were not available by the fact that generally method used for determination content of this element in milk are not enough accurate, because of detection limit of 0.005 ppm. Caggiano et al. (2005) found 0.0025 ppm mercury in sheep milk, and Saitmuratova et al. (2001) found 0.01 ppm mercury in camel milk.

The aim of this study is to establish a first diagnostic of contamination state by heavy metals on fodder, water and camel milk in several region of Kazakhstan, and to assess the links between heavy metals content in water, in fodder and in camel milk.

2. Materials and Methods

2.1. SAMPLING PROCEDURE

Eight farms were sampled for water, fodder and camel milk. The samples correspond to water and fodder consumed by camels. They belonged to four regions: Almaty, Aral, Atyrau, Chymket. Samples were taken from dromedary camels (one hump), Bactrian camels (two humps) and hybrids camels. In some farms were other species were reared, milk samples were also collected (mare, cow).

The contaminating sources, as manufacturing, oil forages, spatial bases were identified. When it was possible water and plants were sampled near these sources.

In Almaty region, polymetal industries situated in Tekeli and Cary-Ozek were identified and sampled. In Atyrau region, oil forage was identified too. In Aral Sea region the spatial base of Baïkanour was also identified. And in Chymket region, polymetal Industries and phosphate manufacturing were identified (Kengtaw, Aca).

The localization of sampled farms and contamination sources are represented in Figure 1. Analysis of heavy metals and trace elements was made by ICP methods, in laboratory of CIRAD (Montpellier-France).

2.2. STATISTICAL PROCEDURE

As few data were available, only descriptive statistics were achieved (mean and standard deviation). In order to compare the concentrations for different trace element and heavy metal in fodder and milk, an index was used. This index was calculated for each type of sample (milk or fodder) as follow:

$$I = x^{ij}/mean(x^i)$$

where x^{ij} was the concentration x for the element i and the sample j

Mean(x^i) was the mean of all concentrations for the element i.

If the concentration for a determined element was equal to the mean, the index will be 1.

Correlation between values in fodder and milk was assessed by the Pearson correlation test.

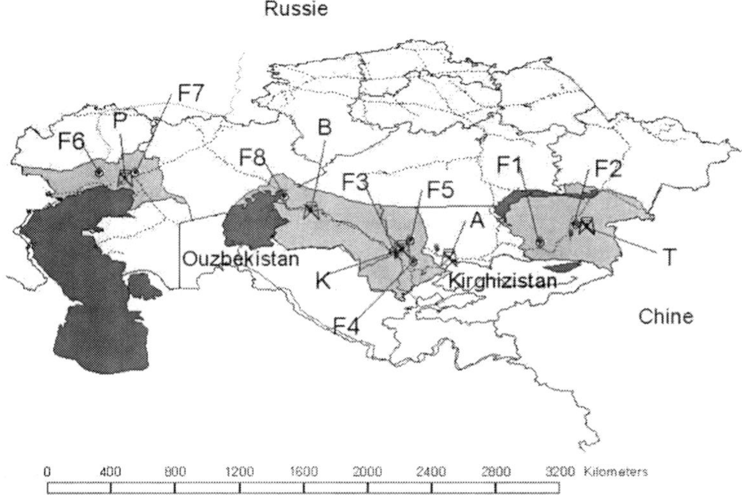

Legend: F1 to F8 = Farm n°1 to Farm n°8 , K = Kengtaw , A = Aca , T = Tekeli , B = Baîkanour , P = oil forage

Figure 1. GPS map of sampling place and localization of pollution sources

3. Results

3.1. TRACE ELEMENTS CONTENTS IN WATER, FODDER AND CAMEL MILK

In water, copper, iron, manganese and zinc were not detected. Concerning trace elements, the maximum value for iron was 2533 ppm, with a high standard error (Table 1). Other elements were in the normal range (Table 1). Copper concentration was lower in camel milk compared to other species. Iron and zinc was strongly higher in mare milk (Table 2).

TABLE 1. Trace element content in fodder (ppm)

	Cu	Fe	Mn	Zn
Mean ± SE	10.40 ± 2.93	793.69 ± 630.48	62.38 ± 20.67	32.95 ± 27.15

TABLE 2. Trace element content in camel milk, and in mare, cow milk (ppm)

	Cu	Fe	Mn	Zn
Mean ± SE (camel milk)	0.07 ± 0.04	1.48 ± 0.53	0.08 ± 0.03	5.16 ± 2.17
Camel milk (farm 2)	0.06	1.77	0.07	3.80
Mare milk (farm 2)	0.10	109	0.66	36
Cow milk (farm 2)	0.49	0.89	0.05	2.30

3.2. HEAVY METALS CONCENTRATION IN WATER, FODDER AND MILK

The samples of water didn't contain lead, and the concentration in arsenic was under analytical limit (less than 0.01 ppm).

On average, samples of fodder contained 4.28 ± 9.60 ppm of lead and 1.03 ± 0.49 ppm of arsenic. The highest value of lead in fodder was 34.90 ppm and was obtained in sample of Kengtaw near a polymetal industry in Almaty region.

In camel milk, the lead concentration was on average 0.025 ± 0.02 ppm, with a maximum value of 0.06 ppm reported in milk sample from the farm n° 4 (Chymket region). The arsenic concentration was under analytical limit (less than 0.1 ppm).

There was no difference between species (mare, cow and camel) in the farm 2 where the three species were reared (Table 3).

TABLE 3. Heavy metal concentration (in ppm) of milk of camel, mare and cow sampling in farm 2 (Almaty region)

	As	Pb
Mare milk	<0.1	0.03
Cow milk	<0.1	0.02
Camel milk	<0.1	0.02

3.3. LINKS BETWEEN TRACE ELEMENT AND HEAVY METALS CONTENT IN FODDER AND IN CAMEL MILK

No clear links between trace element concentration in fodder and milk or between heavy metals concentration in fodder and milk were unclear.

For **copper** (Figure 2), two groups of farms were identified: one group with index in milk lower than index in fodder, and an other group with increasing index from fodder to milk. Two farms showed extreme tendency, the farm 8 (Aral region) with a high index in fodder and the lower index for milk, and the farm 3 (Chymket region) with a low index for fodder and the higher index for milk.

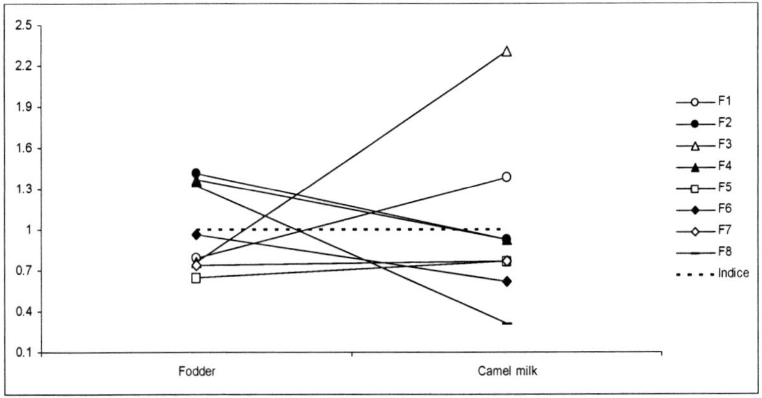

Figure 2. Index of copper content in fodder and in camel milk in each farm sampled

As for copper, two groups for **iron** content were identified (Figure 3). The farm 4 (Chymket region) showed a very high index for fodder (3.6 times than the index 1) and a low index for milk (under the index 1).

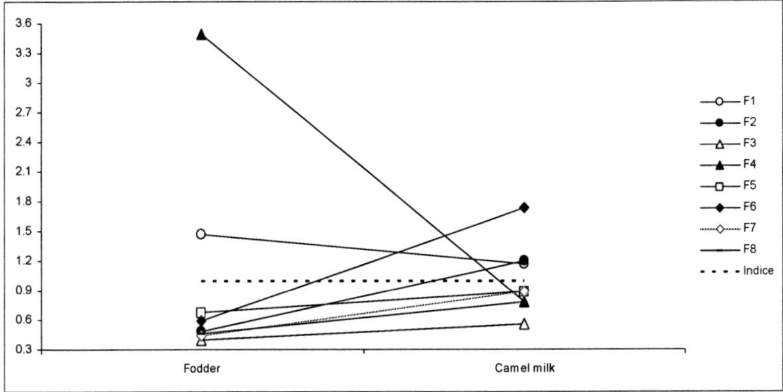

Figure 3. Index of iron content in fodder and in camel milk in each farm sampled

Similar trends were observed for **zinc** (Figure 4) with two groups of farms. All the milk samples had the index under 1, except in farm 8 where the zinc concentration in milk was especially high (Figure 4).

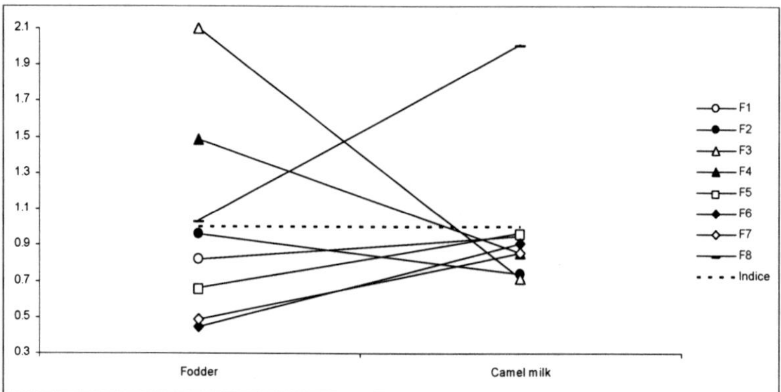

Figure 4. Index of zinc content in fodder and in camel milk in each farm sampled

In milk, the analytical method used didn't allow to determinate precisely the **arsenic** concentration. So, the link between arsenic content in fodder and in milk could not be determinate (Figure 5).

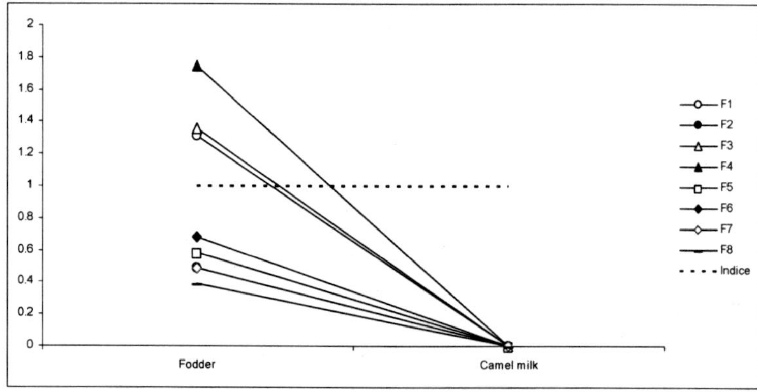

Figure 5. Index of arsenic content in fodder and in camel milk in each farm sampled

Two groups of farms were also identified according to **lead** concentration in fodder and milk samples: one group with a decrease of the index between fodder and milk, and another group with an increase of the index (Figure 6). Lead is not a natural element in fodder and in milk.

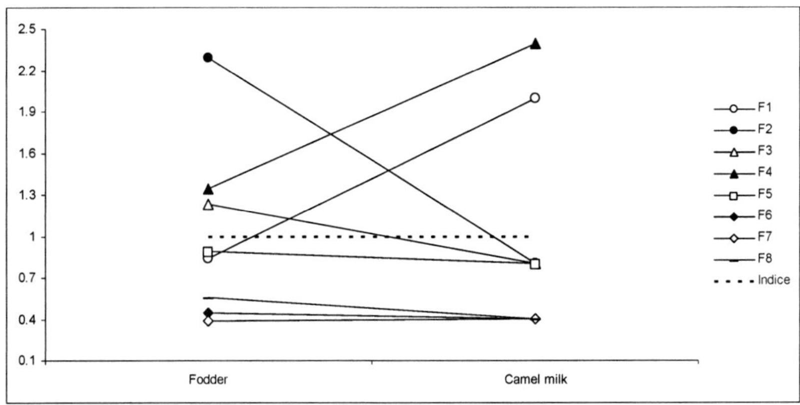

Figure 6. Index of lead content in fodder and in camel milk in each farm sampled

As the whole, there was no correlation between the content of one element (trace element and heavy metal) in fodder and the content of this same element in camel milk. However, the copper and manganese concentration in milk was significantly correlated with zinc content in fodder ($r = 0.724$ and $r = -0.725$), and lead concentration in milk was significantly correlated with iron and manganese content in fodder ($r = 0.897$, $r = 0.815$).

TABLE 4. Element trace and heavy metal concentration (in ppm) in fodder from source of pollution

	Cu	Fe	Mn	Zn	As	Pb
Fodder Kengtaw (SPK)	11.20	636	82.10	94.30	1.60	34.90
Fodder Cary-Ozek (SPCo)	12.50	709	58.90	78.80	0.80	2.60
Fodder farm 3	7.9	291	37.8	52.60	1.40	2.20
Fodder farm 4	14.2	2533	112	37.20	1.80	2.40
Fodder farm 2	14.7	350	38.2	24.00	0.50	4.10

Fodder from the pollutant source in Kengtaw (polymetal industry) was very rich in lead (Table 4). Fodder from Cary Ozek, was richer in copper than fodder from Kengtaw, and the fodder from the farm 2 was richer in copper than the samples from the other farms (F2 was at 64 km from Cary-Ozek). The farms 3 and 4 were close to Kengtaw (42 and 93 km). As for the fodder sampled in Kengtaw compared to fodder sampled in Cary-Ozek, samples from the farm 3 contained more zinc than the others farms, and the farm 4 more manganese and arsenic. The milk sampled in the farm 4 was richer in lead (0.06 ppm) than milk from the others farms.

4. Discussion

4.1. TRACE ELEMENTS AND HEAVY METALS IN FODDER

In this study, trace element concentrations in fodder samples were in the normal range of values. The sample from the farm 4 (Chymket region) contained a quite higher iron quantity (2,533 ppm) compared to the other samples. Iron content in fodder and grass was generally highly variable because the contamination of samples could occur easily.

The manganese content in fodder (62.38 ppm) was in agreement with the results reported by Caggiano et al. (2005) in Italia, in areas exposed to some industrial contamination (101 ± 85 ppm).

Concerning the lead content in fodder sample from Kengtaw (source of pollution, polymetal manufacturing), a very high concentration (34.9 ppm) was reported. On average, the fodder samples analyzed for this study presented an important lead level (4.28 ± 9.60 ppm), higher than Caggiano et al. (2005)'s result (1.2 ± 1.1 ppm). According to the standard error, a high variability was observed. This variability in lead concentration could be attributed to the different risk level in contamination according to the place of sampling.

4.2. TRACE ELEMENTS AND HEAVY METALS IN MILK

On average the trace element concentrations in camel milk were in the normal range reported in the literature. The mean copper concentration (0.07 ± 0.04 ppm) was lower than observations reported in cow milk (Simsek et al., 2000) in industrial area (0.96 ppm) and in traffic area (0.58 ppm). The iron content (1.48 ± 0.53 ppm) was in agreement with other studies on camel milk, but our values were lower than those reported by others authors on polluted areas. For manganese concentration, our results (0.2 ppm) presented were lower than those reported by Sawaya et al. (1984) on cow milk (0.08 ± 0.03 ppm on average). Camel milk content in zinc was in accordance with the literature's results. Sawaya et al. (1984) and Al-Awadi and Srikumar (2001) found respectively 4.4 ± 0.4 ppm and 4.9 ± 0.5 ppm of zinc in camel milk. But regarding result of Meldebekova (unpublished results) on camel milk in Kazakhstan (1.288 ± 1.256 ppm), our value (5.16 ± 2.17 ppm) was higher on average. The farm 8 (Aral Sea region) contained a high level of zinc compared to the others samples (10.4 ppm). Regarding the survey of Simsek et al. (2000) in industrial and in traffic areas (5.01 ppm and 4.49 ppm respectively), the present values were in the same range. These results were also in the normal range of zinc content in milk reported by FAO and consequently seemed to be not in excess.

Arsenic content in camel milk in this study was not determinate precisely (<0.1 ppm). It was in accordance with literature. Simsek et al. (2000), Licata et al. (2004) and Meldebekova et al. (2007) reported values of arsenic concentration in milk less than 0.1 ppm (0.050 ppm, 0.0379 ppm and 0.0218 ± 0.057 ppm respectively).

It was impossible to attest if lead contamination of camel milk was important or not. However, the presence of lead in fodder or milk is not normal, and consequently contamination could be expected. Reported values in our study were on average upper than the tolerable value of 0.02 ppm proposed in Turkey, but under the normal value of 0.05 ppm considered in Germany and Holland and 0.1 ppm in Kazakhstan. However, two of our monitored farms showed quite higher lead content in camel milk, the farm 1 (Almaty region) and the farm 4 (Chymket region): respectively 0.05 ppm and 0.06 ppm.

Bhatia and Choudhry (1996), Dey and Swarup (1996) and Simsek et al. (2000) considered that cattle reared close to manufacturing or roads produced a milk containing significantly higher levels of lead. However, in all these studies, lead concentration in milk was very variable, with higher value between 0.032 ppm (Simsek et al., 2000) and 7.20 ppm (Bhatia and Choudhri, 1996). Taking in consideration all these informations, it was possible that the highest lead concentration in milk from the farms 1 and 4 could be due to the proximity of road with high traffic.

According to the species, the milk composition in trace elements seemed highly variable Indeed, the milk samples of mare, cow and camel from the farm 2 showed differences in trace element concentration with the same mineral composition in food intake. These differences are probably due to different absorption, accumulation and excretion process between species. As it was enlightened, mare milk was very rich in iron, manganese and zinc towards cow and camel milk. And cow milk was richer in copper. For heavy metal concentration in mare, cow and camel milk, it seemed to have no differences. But, in this case study, it was difficult to assess the difference in heavy metal concentration between the three dairy species, because of few numbers of samples. It would be interesting to investigate this aspect with more samples around a pollution source. By this way, it would be possible to assess if there were differences in metabolization, accumulation and excretion of heavy metals in milk for a same contamination level.

4.3. LINKS BETWEEN TRACE ELEMENT AND HEAVY METALS CONTENT IN FODDER AND CONCENTRATION IN CAMEL MILK

No references were available on this aspect. All investigations on contamination of milk by heavy metals, didn't take in account the intake of contaminants by the feeding. So, in this case there was no point of reference for the discussion of the results presented here. The links between trace element and heavy metal content in fodder and concentration in milk were unclear especially because the few number of analyzed samples..

No correlation between fodder and milk content of heavy metals and trace elements was observed in our study. However, an effect of distance from the pollution source of contamination on lead concentration in milk (case of farms 1 and 4), and on fodder content of zinc, manganese, arsenic and lead seemed to be observed. But no references were available in the literature.

Our results were not sufficient to attest the real link between content of heavy metals in fodder and concentration of heavy metals in camel milk.

5. Conclusion

The present investigation on state of contamination of fodder and camel milk by heavy metals in Kazakhstan was a preliminary study. Although, the relationships between heavy metals content in water, fodder and milk were not clear in this study. Some informations were lacking, as references, standard, environmental conditions. Secondly, the number of samples was not sufficient.

It would be interesting to extend the milk and fodder sampling to more areas where camel, cow, goat and sheep farms were closed to pollution areas, and to analyze other heavy metals suspected in pollution process.

References

Al-Awadi, F.M., Srikumar, T.S., 2001, Trace elements and their distribution in protein fractions of camel milk in comparison to other commonly consumed milks, *J. Dairy Res.*, 64: 463–469.

Bhatia, I., Choudhri, G.N., 1996, Lead poisoning of milk—The basic need for the foundation of human civilization, *Indian J. Public Health*, 40(1): 24–26.

Caggiano, R., Sabia, S., D'Emilio, M., Macchiato, M., Anastasio, A., Ragosta, M., Paino, S., 2005, Metal levels in fodder, milk, dairy products, and tissues sampled in ovine farms of southern Italy, *Environ. Res.*, 9: 48–57.

Dey, S., Swarup, D., 1996, Lead concentration in bovine milk in India, *Arch. Environ. Health*, 51(6): 478–479.

Fursov, V.I., Sofrygina, M.T., Romanova, E.V, 1985, Content of heavy metals in fresh cow milk and animal feed. In "Hygiene of environment". In Russian "Gigiena okruzhaiushchey sredy". Ed. Ministry of health. pp. 39–42.

Licata, P., Trombetta, D., Cristani, M., Giofrè, F., Martino, D., Calò, M., Naccari, F., 2003, Levels of "toxic" and "essentials" metals in samples of bovine milk from various dairy farms in Calabria, Italy, *Environ. Int.*, 30: 1–6.

Meldebekoba, A.A., Konuspayeva, G.S., Kaldibekova, H.B., Faye, B., 2007, *Heavy metals residues in camel milk (in Russian)*. Review KazGU, Coll. Ecology, 7 p.

Saitmuratova, O.Kh., Sulaimanova, G.I., Sadykov, A.A., 2001, Camel's milk and shubat from the aral region, *Chem. Nat. Comp.*, 37(6): 566–568.

Sawaya, W.N., Khalil, J.K., Al-Shalhat, A., Al-Mohammad, H., 1984, Chemical composition and nutritional quality of camel milk, *J. Food Sci.*, 49: 744–747.

Simsek, O., Gültekin, R., Öksüz, O., Kurultay, S., 2000, The effect of environmental pollution on heavy metal content of raw milk, *Nahrung*, 44: 360–363.

Swarup, D., Patra, R.C., Naresh, R., Kumar, P., Shekhar, P., 2005, Blood lead levels in lactating cows reared around polluted localities, transfer of lead into milk, *Sci. Total Environ.*, 349: 67–71.

Tripathi, R.M., Raghunath, R., Sastry, V.N., Krishnamoorthy, T.M., 1999, Daily intake of heavy metals by infants through milk and milk products, *Sci. Total Environ.*, 227: 229–235.

Vidovic, M., Sadibasic, A., Cupic, S., Lausevie, M., 2005, Cd and Zn in atmospheric deposit, soil, wheat, and milk, *Environ. Res.*, 97: 26–31.

HEAVY METALS AND TRACE ELEMENTS CONTENT IN CAMEL MILK AND *SHUBAT* FROM KAZAKHSTAN

ALIYA MELDEBEKOVA, GAUKHAR KONUSPAYEVA
Kazakh State University Al Farabi, Almaty, Kazakhstan

EMILIE DIACONO, BERNARD FAYE
Département environnement et société, CIRAD, Montpellier, France

Abstract: In Kazakhstan, camel milk is mainly consumed after fermentation process. The fermented camel milk, named *shubat*, is generally home-made by the traditional process. The changes in mineral composition of camel milk during the fermentation process were rarely studied especially for heavy metals. The present study aimed to assess the change in heavy metals and trace-elements contents during the fermentation process.

Samples of milk and *shubat* were collected in eight farms of Southern Kazakhstan in order to determine copper, iron, manganese, zinc, arsenic and lead. In camel milk mean content of these heavy metals was respectively of 0.065 ± 0.04, 1.478 ± 0.53, 0.084 ± 0.03, 5.163 ± 2.17, <0.1 and 0.025 ± 0.02 ppm. In *shubat*, the mean content was 0.163 ± 0.164, 1.57 ± 0.46, 0.088 ± 0.02, 7.217 ± 2.55, and 0.007 ppm respectively.

Arsenic was detected in some samples of milk and *shubat* only. A relationship between heavy metals in raw milk and *shubat* at the farm level was observed.

Keywords: Camel milk, camel milk products, pollution, heavy metals, trace element

1. Introduction

Kazakhstan is a country where the milk consumption is high since it reaches 240 l/hab/year most of the time in liquid form. Another characteristic of this country is to have on the one hand important milk coming from non-conventional species like the mare and the camel one. These milks are generally consumed in fermented form (named *kumis* and *shubat*). They give culturally important products taking significant part in the Kazakh identity.

Shubat has the reputation to have medicinal and probiotic properties largely made profitable by the medical profession (Sinyavskiy, 2004), particularly in the Sanatorium where fermented milks are used as additive in the tuberculosis treatment as well in the countries of the ex-Soviet Union (Kadyrova, 1985; Kenzhebulat et al., 2000) as in the countries of the South (Mal et al., 2006). The camel milk is considered also to have antidiabetic properties (Agrawal et al., 2003), anti-cancer (Magjeed, 2005), and more generally to have dietetic quality because of its richness in unsaturated fatty acids (Narmuratova et al., 2006; Karray et al., 2005). One allots even anti-contaminant properties with respect to the radionuclide. All these properties are however generally based on empirical observations or protocols of insufficiently rigorous observation.

According to the high polluting conditions of some parts of the environment In Kazakhstan, camel milk and *shubat* could contain also pollutants (pesticides, heavy metals, radionuclides). However, as *shubat* is the main product consumed by human, the assessment of the transfer in pollutants from raw milk to *shubat* is an important point to study. So, the present paper is focused on the mineral components of raw milk and *shubat* in different farms from Kazakhstan situated close to pollution sources in order to assess the risk for human consumers.

B. Faye and Y. Sinyavskiy (eds.), *Impact of Pollution on Animal Products.*
© Springer Science + Business Media B.V. 2008

2. Materials and Methods

2.1. SAMPLING PROCEDURE

Camel milk and *shubat* samples were collected from 8 farms situated in four regions in the south-eastern, southern and the western part of Kazakhstan (Figure 1). The sampling procedure was carried out following way: during the hand milking procedure of all milked animals in a corral the milk was collected in big dishes with capacity of 50 l. It was taken one sample of milk from this mix immediately after the finish of milking and one sample of fermented milk at each farm. *Shubat* was prepared from mixed milk from all milked camels with addition of ready (strong) *shubat* as ferment in a definite mixing with raw milk. All samples in plastic bottles were put in portable freeze with icing elements (pockets) inside for transportation and in further kept frozen until the analyses. Samples were analyzed for copper (Cu), manganese (Mn), iron (Fe), zinc (Zn), arsenic (As) and lead (Pb) content.

Figure 1. Sampling sites of camel milk and *shubat* in Kazakhstan, F1–F8 – farms

2.2. LABORATORY ANALYSIS

Analyses included two procedures – the dry way mineralization and silica elimination by fluorhydric acid (HF) and the measurement of elements. The emission plasma spectrometry (inductively coupled plasma – ICP) method used with apparatus Varian-Vista for determination elements' content in samples. All analyses were provided in certified Laboratory (ISO 9001 quality control) of physic-chemical analyses of plant, soil and water at CIRAD (Montpellier, France) with internal reference samples containing already known elements rate.

2.3. STATISTICAL ANALYSIS

Only descriptive analysis with XLstat software was used for data analysis. Data were presented by the mean ± standard-deviation. As only one sample was achieved, no statistical test was used.

3. Results

3.1. MINERALS CONTENT IN RAW MILK

The mean values for milk were 0.065, 1.478, 0.084, 5.163, <0.1 and 0.025 ppm for Cu, Fe, Mn, Zn, As and Pb respectively (Table 1).

TABLE 1. Mean, standard deviation, maximum and minimum values for minerals and heavy metals in camel milk

Element (ppm)	n	$\mu \pm SD$	Max	Min
Cu	8	0.065 ± 0.04	0.15	0.02
Fe	8	1.478 ± 0.534	2.56	0.83
Mn	8	0.084 ± 0.03	0.14	0.04
Zn	8	5.163 ± 2.168	10.40	3.70
As	8	<0.1	<0.1	<0.1
Pb	8	0.025 ± 0.019	0.06	0.01

3.2. MINERALS CONTENT IN *SHUBAT*

Concerning *shubat*, the mean values for Cu, Fe, Mn, Zn, As and Pb respectively, were 0.163, 1.57, 0.088, 7.217, <0.1 and 0.007 ppm (Table 2).

TABLE 2. Mean, standard deviation, maximum and minimum values for minerals and heavy metals in *shubat*

Element (ppm)	n	$\mu \pm SD$	Max	Min
Cu	6	0.163 ± 0.164	0.47	0.02
Fe	6	1.57 ± 0.46	2.48	1.16
Mn	6	0.088 ± 0.018	0.1	0.06
Zn	6	7.217 ± 2.555	11.8	4.9
As	6	<0.1	<0.1	<0.1
Pb	6	0.007 ± 0.005	0.01	0.0

3.3. RELATIONSHIPS BETWEEN RAW MILK AND SHUBAT MINERAL CONTENTS

The relationship in mineral content of raw camel milk and *shubat* was observed on 6 farms only because *shubat* sample lack from the farms 1 and 6. Concerning **Cu** content, generally, mean value of this mineral in *shubat* was higher than mean for raw milk. According to the results, farms could be divided into two types. In fact, in samples from farms 3, 2, 7 and 5 its concentration increased in *shubat*, but in farms 4 and 8 the difference on its content between types of product was a few (Figure 1). The highest number detected in F3 *shubat* – 0.47 ppm, while the mean is 0.16 ppm.

The mean **Fe** value for *shubat* is also higher than that for milk, in most of the case except in F2 and F7 farms. The highest number was detected in *shubat* of F8 (2.48 ppm) from Aralsk zone (Figure 2).

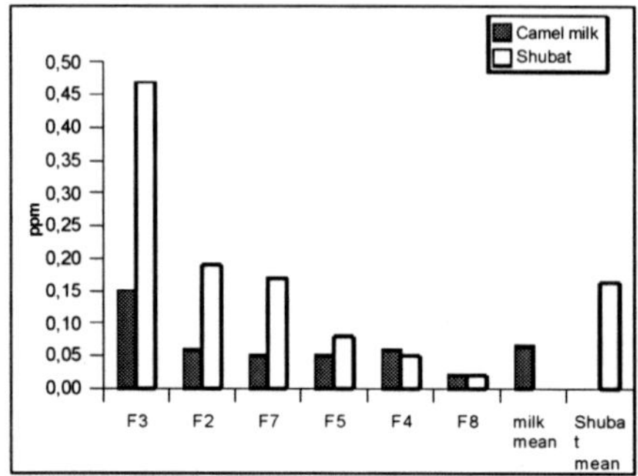

Figure 1. Cu content in camel milk and *shubat*

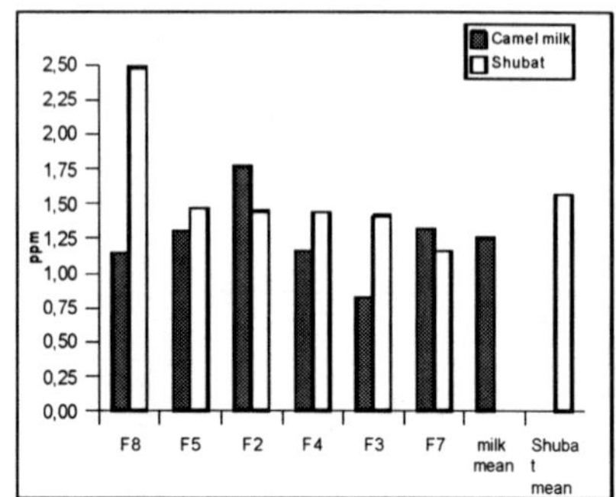

Figure 2. Fe content in camel milk and *shubat*

Farms weren't differed noticeable by **Mn** content, but milk from F7 showed highest level (0.14 ppm) compared to other samples (Figure 3). *Shubat* samples from F2, F3 and F4 contained more Mn than milk.

Zn content increased in all *shubat* samples (except F8). It should be noticed that Zn content in *shubat* from F3 was rather high (11.8 ppm), whereas the value for the milk from the same farm was less in about threefold (3.7 ppm) (Figure 4).

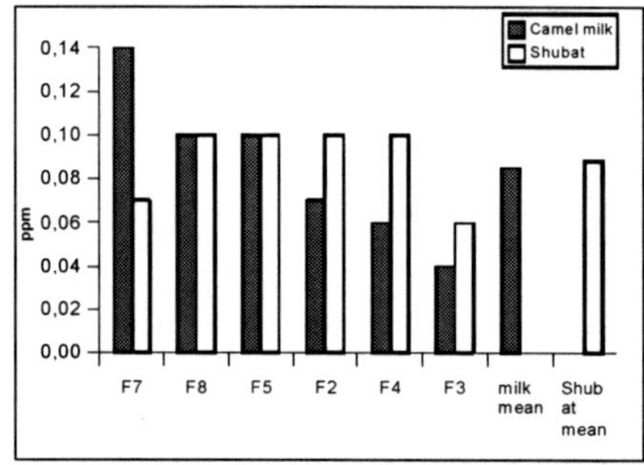

Figure 3. Mn content in camel milk and *shubat*

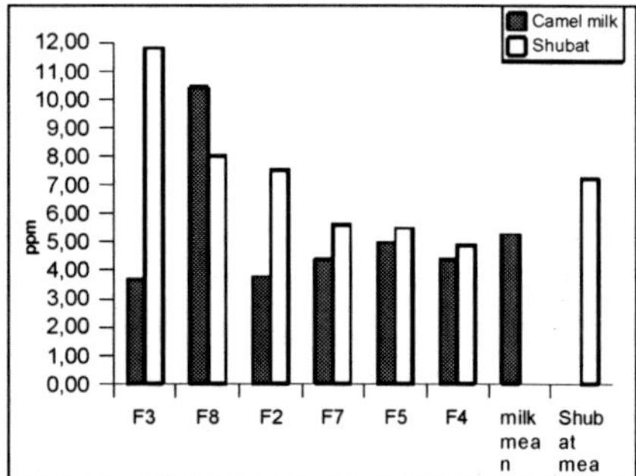

Figure 4. Zn content in camel milk and *shubat*

In another hand, the twofold high value (10.4 ppm) in comparison with average (5.3 ppm) for milk at F8 decreased in *shubat* (8.0 ppm).

Concerning **Pb**, as a whole, its content decreased in *shubat*, particularly, Pb was not detected in *shubat*s of F4 and F5, although the maximal value is detected in milk of F4 (0.06 ppm) (Figure 5).

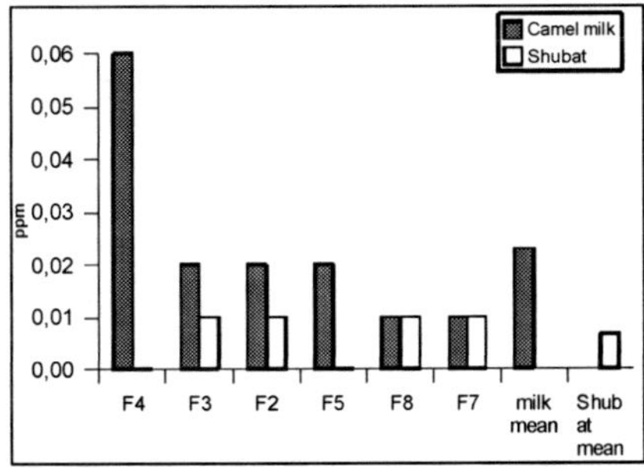

Figure 5. Pb content in camel milk and *shubat*

3.4. RESULTS FROM DIFFERENT LABORATORIES

It was investigated the only camel milk sample from the farm 1 (Almaty region) in 3 different laboratories for minerals mentioned above. It was used various methods of analysis: in the Lab1 – ICP method, in the Lab2 – Volt-amperometric (potentiometric) and in the Lab3 – Atomic absorbtion spectrometric (AAS-3) methods. Consequently, some differences were found in results (Figure 6). In particular, Fe and Zn content lower (0.46 and 0.4 ppm) according to the Lab2, when by Lab1 it was 1.72 (Fe) and 4.9 (Zn) ppm (Figure 6). At the same time Cu content was higher according to the Labs 2 and 3 (0.51 and 0.27 ppm) *vs* Lab1 (0.09 ppm) (Figure 6.2). The As content remained not exactly clear: it was 0.02 and 0.004 ppm by Labs 2 and 3. The Pb content was 0.05, 0.069 and 0.01 ppm in respect with labs.

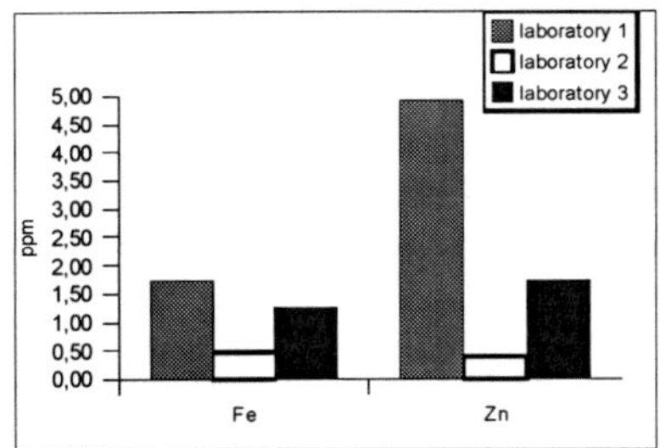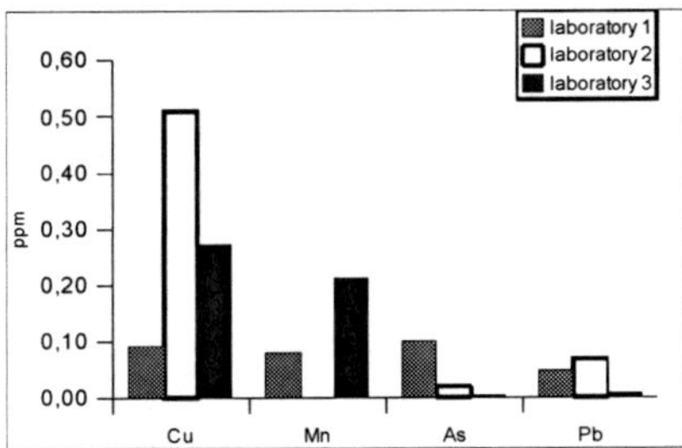

Figure 6. Content of Fe, Zn (1), Cu, Mn, As and Pb (2) in a sample of camel milk by different laboratories

4. Discussion

The Cu value according to our results (0.065 and 0.163 ppm for milk and *shubat*) was lower than in literature data on cow milk (Simsek et al., 2000; Campilo et al., 1998; cited by Licata et al., 2004) but within the range of normal values. The normal range of its content in cow's milk is 0.1–0.4 ppm (FAO) and 1.0 ppm for camel milk and *shubat* in Kazakhstan.

Concerning Fe, the essential micro-element of the milk, camel milk is rich in it compared to other livestock species, according to literature (Konuspayeva, 2007). Our iron values of camel milk (1.47 ppm) and *shubat* (1.57 ppm) seem to be lower than in some literature data (Sawaya et al., 1984; Bengoumi et al., 1998). Some references available reporting high iron contents: 32 and 77 Fe ppm for camel milk and *shubat* respectively (Saitmuratova et al., 2001).

The mean Mn concentration in 0.08 ppm for both of our milk and *shubat* are lower compared to mean results on sheep – 0.13 (Caggiano et al., 2005) and 0.31 ppm (Coni et al., 1996; cited by Caggiano et al., 2005) and goat milk – 0.13 ppm (Coni et al., 1996 cited by Caggiano et al., 2005). No references are available on the normal value of it in camel milk.

Concerning Zn content in our samples (5.1 and 7.2 ppm on average), it seems to be more close to Simsek results on cow milk: 5.01 ppm in industrial area, 4.49 ppm in intensive traffic area and 3.77 ppm in rural area (Simsek et al., 2000). The normal rate for it is from 3 up to 6 ppm (FAO) and 5.0 ppm for camel milk and *shubat* in Kazakhstan. Maximum values in our study (11.8 ppm for *shubat* from Turkestan city and 10.4 ppm for milk from *Aral* Sea zone) are rather high. Saitmuratova et al. reported high values (59 and 33 Zn ppm for camel milk and *shubat*) on this element too (Saitmuratova et al., 2001), and their samples were from that zone, also.

The As detected in concentration less than 0.1 ppm only, without exact number. No data were available on its concentration in camel milk but according to the Kazakh standard camel milk and *shubat* should not contain more than 0.055 as ppm.

Pb mean concentrations (0.025 and 0.007 ppm) were around the value of 0.02 ppm in Turkey, under the acceptable value of 0.05 in Germany, Holland and 0.1 ppm in Kazakhstan. Pb value in milk samples from F1 *(Almaty region)* and F4 *(Shymkent region)* were higher (0.05 and 0.06 ppm), but in *shubat* its decreased sharply.

5. Conclusion

As a whole, there was not detected trace elements and heavy metals pollution according to our results. Only the Zn mean *shubat* content seems to be slightly high according to the national standard for *shubat*. A *shubat* sample from F3 *(Turkestan)* had highest content of Cu and Zn than in other farms, although, it should be more expected in milk than *shubat* because of farm situating not far away from the formerly and/or still working Cu and Zn mining plants in *Turkestan* and *Kentau* cities. The Pb content in *shubat* decreased in comparison with camel milk. So, the link between trace elements and heavy metals changing in camel milk and *shubat* remained not clear in present study.

The differences in results from different methods points to the necessity for centralisation and standardisation of methods used for analyses. In summary, all these facts point to the necessity on providing more detailed studies in this field with taking in account more potential risk elements and factors.

References

Agrawal, R.P., Swami, S.C., Beniwal, R., Kochar, D.K., Sahani, M.S., Tuteja, F.C., Ghouri, S.K., 2003, Effect of camel milk on glycemic control, risk factors and diabetes quality of life in type-1 diabetes: a randomised prospective controlled study, *J. Camel Res. Pract.*, 10: 45–50.

Bengoumi, M., Faye, B., Tressol, J.C., 1998, Composition minérale du lait de chamelle du sud marocain. Pages 145–149 in Actes de l'atelier "Chameaux et dromadaires, animaux laitiers", Nouakchott, Mauritania, 24–26 October 1994, P. Bonnet (ed.). CIRAD, Montpellier, France.

Caggiano, R., Sabia, S., D'Emilio, M., Macchiato, M., Anastasio, A., Ragosta, M., Paino, S., 2005, Metal levels in fodder, milk, dairy products, and tissues sampled in ovine farms of southern Italy, *Environ. Res.*, 99: 48–57.

Campilo, N., Vinas, P., Lòpez-Garcìa, L., Hemàndez-Còrdoba, M., 1998, Direct determination of copper and zinc in cow milk, human milk and infant formula samples using electrothermal atomization atomic absorption spectrometry, *Talanta*, 46: 615–622.

Coni, E., Bocca, A., Coppolelli, P., Caroli, S., Cavalluci, C., Trabalza Marinucci, M., 1996, Minor and trace element content in sheep and goat milk and dairy products, *Food. Chem.*, 57: 253–260.

Kadyrova, R.X., 1985, Le lait de chamelle et le lait de jument comme alimentation à visée médicale, publ. Alma-Ata, Kazakhstan, 158 p.

Karray, N., Lopez, C., Ollivon, M., Attia, H., 2005, La matière grasse du lait de dromadaire: composition, microstructure et polymorphisme. Une revue, *OCL*, 12: 439–446.

Kenzhebulat, S., Ermuhan, B., Tleuov, A., 2000, Composition of camel milk and it's use in the treatment of infectious diseases in human. In "2nd Camelid Conf. Agroeconomics of camelid farming", AgroMerkur Publ., Almaty, Kazakhstan, 8–12 September 2000, 101.

Konuspayeva, G., 2007, Variabilité physico-chimique et biochimique du lait des grands camélidés (Camelus bactrianus, Camelus dromedarius et hybrides) au Kazakhstan. Thèse en Sciences des aliments. Université de Montpellier II (France).

Licata, P., Trombetta, D., Cristani, M., Giofrè, F., Martino, D., Calò, M., Naccari, F., 2004, Levels of "toxic" and "essentials" metals in samples of bovine milk from various dairy farms in Calabria, Italy, *Environ. Int.*, 30: 1–6.

Magjeed, N.A., 2005, Corrective effect of camel milk on some cancer biomarkers in blood of rats intoxicated with aflatoxin B1, *J. Saudi Chem. Soc.*, 9(2): 253–263.

Mal, G., Sena, D.S., Jain, V.K., Sahani, M.S., 2006, Therapeutic value of camel milk as a nutritional supplement for multiple drug resistant (MDR) tuberculosis patients, *Israel J. Vet. Med.*, 61: 88–91.

Narmuratova, M., Konuspayeva, G., Loiseau, G., Serikbayeva, A., Barouh, N., Montet, D., Faye, B., 2006, Fatty acids composition of dromedary Bactrian camel milk in Kazakhstan, *J. Camel Pract. Res.*, 13: 45–50.

Saitmuratova, O.Kh., Sulaimanova, G.I., Sadykov, A.A., 2001, Camel's milk and shubat from the aral region, *Chem. Nat. Comp.*, 37(6): 566–568.

Sawaya, W.N., Khalil, J.K., Al-Shalhat, M., Al-Mohammed, H., 1984, Chemical composition and nutritional quality of camel milk, *J. Food Sci.*, 49: 744–747.

Simsek, O., Gültekin, R., Öksüz, O., Kurultay, S., 2000, The effect of environmental pollution on heavy metal content of raw milk, *Nahrung*, 44: 360–363.

Sinyavskiy, Y.A., 2004, Development of products for child nutrition and for medical and prevention purposes on the base of camel milk, Proc. of Intern. Workshop, "Desertification combat and food safety: the added value of camel producers". Ashkhabad (Turkmenistan), 19–22 April 2004, In "Vol. 362 NATO Sciences Series, Life and Behavioural Sciences", B. Faye and P. Esenov (Eds). IOS press, Amsterdam, The Netherlands, 194–199.

VARIATION FACTORS OF SOME MINERALS IN CAMEL MILK

GAUKHAR KONUSPAYEVA, MIRAMKUL NARMURATOVA,
ALIYA MELDEBEKOVA
Kazakh State University Al Farabi, Almaty, Kazakhstan

BERNARD FAYE, GERARD LOISEAU
Département environnement et société, CIRAD, Montpellier, France

Abstract: In four regions of Kazakhstan (Atyrau, Aralsk, Shymkent and Almaty), a survey on camel farms was achieved in order to study the variability of the physico-chemical composition of camel milk both in dromedary (*Camelus dromedarius*) and Bactrian (*Camelus bactrianus*) camel as well as their hybrids. As the whole, 163 milk samples were analyzed for calcium, phosphorus and iron determination. In order to maximize the variance, the samples were done in four different seasons which expressed the feeding change and the physiological stage changes as the calving season was concentrated in few months. The mean values were respectively 1.232 ± 0.292 g/l, 1.003 ± 0.217 g/l and 2.02 ± 1.24 mg/l for calcium, phosphorus and iron. No species, season or region effect was observed on iron content in the milk. Calcium and phosphorus change significantly according to season and species, but only phosphorus was linked to region effect. Especially phosphorus content is high in Aralsk region (1.156 ± 0.279 g/l). Globally, it is noticeable to observe the high level of phosphorus in the camel milk of Kazakhstan compared to the literature's results.

Keywords: Camel milk, minerals, lead, phosphorus, calcium, iron

1. Introduction

The consumption of camel milk, especially under fermented form (named *shubat*), is a very old tradition in Kazakhstan. Peoples give many beneficial properties to these milk products. It is traditionally used for tuberculosis treatment, in gastro-enteritis, any infectious and also like tonic drunk (Sharmanov and Dzhangabylov, 1991). These medical properties of camel milk and *shubat* could be attributed to some substances, like proteins, lipids and vitamins (Elagamy, 1992, 2000; Farah, 1993; Benkerroum et al., 2004; Konuspayeva et al., 2004). Yet, camel milk is well known also for its richness in minerals (Farah, 1993) interacting with the above substances. It could also contain undesirable metals in case of polluted environment. Indeed, the ecological conditions in Kazakhstan could have an important impact on the milk composition, especially its mineral content due to heavy metals in soil and plants intended for animal feeding.

The current camel population in Kazakhstan is around 145,000 heads (Anonymous, 2006). In the country, the genus *Camelus* included two species cohabiting in the same areas and even on the same farms: the one-humped camel (*Camelus dromedarius*) and the Bactrian two-humped camel (*Camelus bactrianus*), and their hybrids (Terenytev, 1975; Konuspayeva and Faye, 2004). This peculiarity fact allowed the comparison of milk composition of those animals reared in similar environments.

In the present study, four minerals were taken in account, i.e. the main minerals interacting with other components of milk (calcium, phosphorus and iron) and lead which is one of the more toxic heavy metal for human consumers.

2. Materials and Methods

2.1. SAMPLING PROCEDURE

In order to get the maximum variability, the dromedary, the Bactrian and the hybrid camel milk was sampled in four whole different regions at the extreme points of Kazakhstan: Almaty, Atyrau, Aralsk and Shymkent (the maximum distance between the different points was more than 3,500 km) and at the four seasons of the year (Table 1). As a whole, 176 milk samples were collected but 164 only were used for quantitative determination of calcium (Ca) and 163 for phosphorus (P) and iron (Fe).

For lead determination, 63 samples were analyzed by taking in account the same variations factors than for the whole samples.

2.2. LABORATORY ANALYSIS

It included two steps: (1) the mineralisation by dry way and silica elimination by fluorhydric acid (HF), (2) the determination of minerals by emission plasma spectrometry (induced coupled Plasma – ICP) with a Varian Vista N.D.

TABLE 1. Sampling design of the camel milk in Kazakhstan

Factors		*Bactrian*	*Dromedary*	*Hybrid*	*Mixed*	*Unknown*	*Total*
Almaty	Winter	3	5				8
	Spring	5	14				19
	Summer	4	15				19
	Autumn	4	2				6
Total Almaty		**16**	**36**				**52**
Atyrau	Winter	7	2				9
	Spring	5	2	1	2		10
	Summer	7	1		1		9
	Autumn	6	1		1		8
Total Atyrau		**25**	**6**	**1**	**4**		**36**
Aralsk	Winter	2	2				4
	Spring	1	2	4	2		9
	Summer	1	1	2			4
	Autumn	1			1		2
Total Aralsk		**5**	**5**	**6**	**3**		**19**
Shymkent	Winter		2	2	1		5
	Spring	3	2	5	7	6	23
	Summer	2	6	5	6		19
	Autumn	1	6	1	2		10
TotalShymkent		**6**	**16**	**13**	**16**	**6**	**57**
TOTAL		**52**	**63**	**20**	**23**	**6**	**164**

All the dosages were achieved after validation with references samples containing known rate in minerals. The analysis was achieved at the agro-food laboratory in CIRAD-Montpellier (France).

2.3. STATISTICAL ANALYSIS

A linear model was tested for Ca, P and Fe concentrations as dependant variables. The tested variation factors were the region, the species, the season and their interactions. The limit of signification level for variance analysis was 0.05. The results are presented as mean plus/minus standard error. As the variances were not homogeneous, data were log transformed in order to get normal distribution of the values. The correlation between the minerals was tested using Pearson correlation. Only descriptive statistics were presented for lead content in milk. In order to identify the discriminating parameters between the species, a discriminant analysis was achieved after eliminating the season and region effect according to the procedure described by Doledec and Chessel (1987). The discriminant analysis included all physico-chemical components of milk described in previous paper (Konuspayeva, 2007). The discriminating power was expressed by the discriminant coefficient (or canonical weight).

The R software was used for all statistical analyses (Ihaka and Gentleman, 1996).

3. Results

3.1. MAIN MINERAL CONTENTS

In camel milk from Kazakhstan, the mean values were 1.232 g/l for calcium, 1.003 g/l for phosphorus and 2.02 mg/l for iron. (Table 2) with a significant effect of season and species for Ca and P. Region effect was observed for P only, but significant interactions occurred between region, season and species (Table 3).

TABLE 2. Mean, standard-deviation, maximum and minimum values for minerals in camel milk

Mineral	n	$\mu \pm SD$	Max	Min
Ca (g/l)	164	1.232 ± 0.292	2.340	0.530
P (g/l)	163	1.003 ± 0.217	1.770	0.520
Fe (mg/l)	163	2.02 ± 1.24	12.40	0.70

TABLE 3. Signification level of the variation factors issued from variance analysis

Mineral	Region	Season	Species	Region*season	Species*season
Ca	0.09	<0.001	0.003	0.14	0.56
P	<0.001	<0.001	<0.001	0.001	0.014
Fe	0.29	0.35	0.55	0.22	0.48

The **calcium** content in milk changed all along the year with similar trends in the different species with a slight increase in spring, a decrease in summer (especially in dromedary and hybrid), then an increase in autumn (except hybrid). The mean values were higher in spring (1.35 ± 0.28 g/l) compared to summer (1.10 ± 029), autumn (1.21 ± 0.22) or even winter (1.23 ± 0.29). The Bactrian milk contained more calcium (1.30 ± 0.29) than dromedary milk (1.16 ± 0.27). Hybrid milk was intermediary (1.26 ± 0.27).

Concerning **phosphorus**, its concentration was higher in Bactrian milk (1.08 ± 0.18 g/l) compared to dromedary milk (0.91 ± 0.19 g/l). The hybrid milk was closed to Bactrian milk (1.07 ± 0.27). Phosphorus content was low in summer (0.86 ± 0.18) and in autumn (0.96 ± 0.14) in all the species compared to spring (1.11 ± 0.22) and winter milk (1.09 ± 0.14). The milk from Aralsk (1.16 ± 0.28) was much higher than Atyrau (1.02 ± 0.16), Almaty (0.99 ± 0.22) and Shymkent (0.95 ± 0.20).

For **iron**, a high variability was observed (0.7–12.4 mg/l) but no variation factor was significant (Table 3).

3.2. CORRELATIONS BETWEEN MINERAL CONTENTS

A significant positive correlation ($r = 0.78$, $P < 0.001$) was observed between Ca and P concentration in milk, both in Bactrian, dromedary and hybrid species (Figure 1).

The ratio Ca/P was 1.23 for the whole samples with a range between 0.57 and 1.93. No significant correlation was reported between Ca and P on one hand and Fe on another hand.

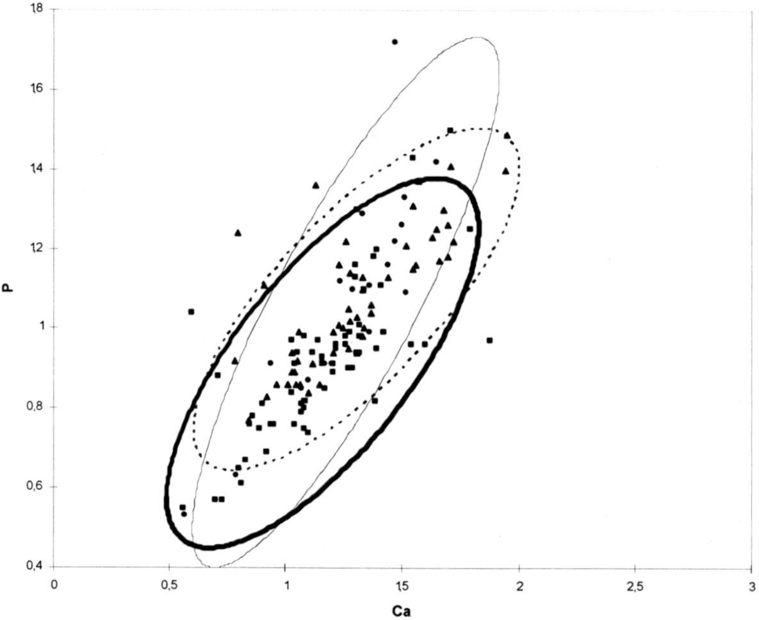

Figure 1. Correlation between Ca and P content in Bactrian (——), dromedary (▬) and hybrid (---) milk

3.3. BETWEEN-SPECIES DISCRIMINANT PARAMETERS

According to the coefficient of linear discriminants (Table 4) issued from simple discriminant analysis, the most discriminant parameter between Bactrian and dromedary milk was the phosphorus concentration (and correlatively the calcium content). The percentage of well classed animals by including six main milk parameters was 75.4%.

TABLE 4. Coefficients of linear discriminant (LD1) between Bactrian and dromedary camel milk issued from simple discriminant analysis

Parameters	LD1
Phosphorus	**−1.00076**
pH	−0.40802
Vitamin C	−0.37738
Iodine index	0.28723
Fat matter	−0.22694
Total protein	−0.00089

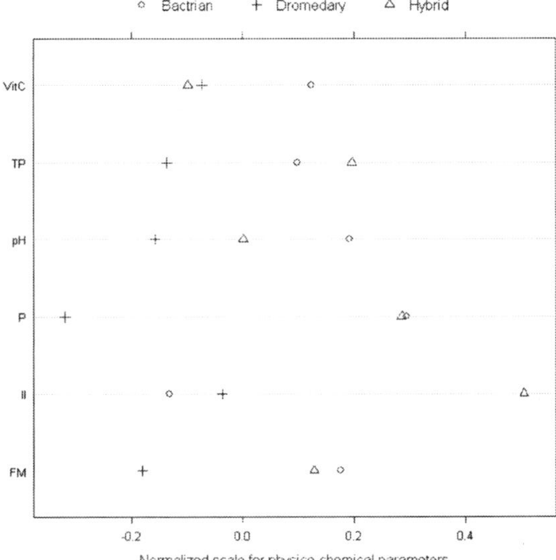

Figure 2. Normalized scale for the physico-chemical parameters (vitamin C, total protein, pH, phosphorus, iodine index, fat matter) of Bactrian milk (o), dromedary (+) and hybrid (⊕) milk samples participating to the discriminant function

The phosphorus and calcium content in milk was indeed significantly higher in Bactrian camel compared to dromedary. When hybrids were included in the discriminant model, phosphorus appeared yet as the most discriminant factor (Figure 2).

3.4. LEAD CONTENT

The lead content was on average 0.250 ± 0.056 ppm in the camel milk with no significant differences between regions: 0.245 ± 0.034 at Almaty, 0.249 ± 0.05 at Aralsk, 0.244 ± 0.083 at Atyrau and 0.260 ± 0.035 at Shymkent. The maximum value was observed at Atyrau (0.532 ppm). There were no differences also between Bactrian (0.233 ± 0.041), dromedary (0.246 ± 0.069), hybrid (0.256 ± 0.027) and mixed (0.286 ± 0.057). However, a seasonal significant effect was observed with a lower lead content in spring (0.212 ± 0.027) compared to winter (0.245 ± 0.043), to summer (0.250 ± 0.061) and to autumn (0.272 ± 0.044).

4. Discussion

On average, mineral matters calculated in 82 references (Konuspayeva, 2007) were in higher quantities in dromedary milk (0.99%) than in references from Central Asia (0.79%) where Bactrian camel was predominant, but few data were available on specific minerals.

If the calcium content in our milk samples was close to the literature's data (Abu-Lehia, 1987; Farah and Ruegg, 1991; Bengoumi et al., 1998) with values between 1.15 and 1.57 g/l, the phosphorus content in camel milk from Kazakhstan appeared in higher concentration than those of the literature: between 0.63 g/l (Sawaya et al., 1984) and 1.04 g/l (Farah and Rüegg, 1989). The higher phosphorus concentration in Bactrian milk compared to dromedary milk could be linked to its higher fat content, especially in phospholipids. As a consequence the ratio Ca/P appeared lower than the literature's data and quite lower than for other species (Figure 3).

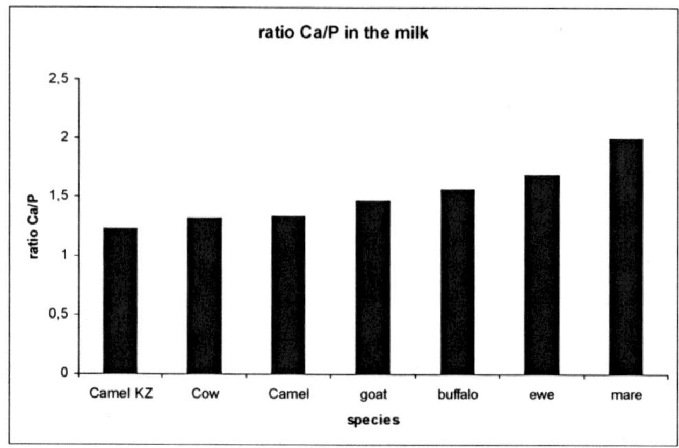

Figure 3. Ca/P ratio in camel milk from Kazakhstan (camel KZ) and other species according to the literature's data

The calcium and phosphorus content in camel milk covered the daily requirements for adult when the consumption reached 650 ml per day of camel milk for calcium and 800 ml per day for phosphorus.

Iron milk content did not discriminate the two camel species. A part of iron is linked to lactoferrin. In a previous publication, no significant difference between species was reported for the lactoferrin content in the milk (Konuspayeva et al., 2007). The observed iron values in our samples (around 2 mg/l) were slightly lower (2.8 to 3.4 mg/l) than those reported by several authors (Sawaya et al., 1984; Abdelrahim, 1987; Bengoumi et al., 1998). In some references, very high iron concentrations were reported: 280 mg/l (Elamin and Wilcox, 1992). The lack of standardized method could explain such differences.

TABLE 5. Iron content in milk from different species (in mg/l) used for human consumers

	Camel KZ	Camel	Cow	Buffalo	Mare	Goat	Ewe
Fe	2.02	2.81	0.35	1.00	0.59	0.55	0.89

Compared to other species, camel milk appeared as very rich in iron (Table 5). However, to cover the daily iron requirements of adults, the consumers have to drink 7.42 l of camel milk for males and 8.91 l for females.

No data was available in literature concerning lead content in camel milk. According to the Kazakh standard, the milk for human cannot overpass 0.1 ppm. It's mean that on average, the camel milk in Kazakhstan contained more than twice the authorized quantity of lead. Some samples appeared as highly contaminated (up to 0.53 ppm). Anyway, all the samples were above the limit of 0.1 ppm. And all regions were affected. In the frame of this study, no link was possible to establish with specific pollution source. Probably, the contamination by fuel from cars and agricultural machines in the sampled farms could be the main origin.

5. Conclusion

The main variation factors (season and region) reflected mainly the nutritional status of the animals and the physiological stage. Indeed, all the animals were reared in extensive system farming with only natural feedstuffs with nutritional values mainly depending on the season. Secondly, the calving season was concentrated within 2 months (February–March), so the seasonal effect included the physiological status of the animals. Calcium and phosphorus were considered as the main milk parameter discriminating dromedary to Bactrian camel milk that it was never described formerly.

The impact of mineral pollution was not detected by the main mineral composition in our samples. However, our preliminary results concerning lead showed high values in all the samples. Other analyses of heavy metals in our samples are in progress (arsenic, zinc, chrome). Those results would be of high importance to assess the risk linked to the regular milk drinking by human consumers.

Acknowledgments

This study was supported by the French Ministry of Foreign Affairs (ECONET Programme) and by the French Embassy at Almaty (Kazakhstan). We also thank the camel farmers from Kazakhstan for their exceptional hospitality.

References

Abdelrahim, A.G., 1987, The chemical composition and nutritional value of camel (*Camelus dromedarius*) and goat (*Capra hircus*) milk, *World Rev. Anim. Prod.*, 18: 9–12.

Abu-Lehia, I.H., 1987, Composition of camel milk, *Milchwissenschaft*, 42: 368–371.

Anonymous, 2006, Short Review of Statistic Agency of Republic of Kazakhstan. 2006. Social-economic development of Kazakhstan, Almaty, pp. 21–22.

Bengoumi, M., Faye, B., Tressol, J.C., 1998, Composition minérale du lait de chamelle du sud marocain. Pages 145–149 in Actes de l'atelier "Chameaux et dromadaires, animaux laitiers", Nouakchott, Mauritania, 24–26 October 1994, P. Bonnet (ed.) CIRAD Montpellier, France.

Benkerroum, N., Mekkaoui, M., Bennani, N., Hidane, K., 2004, Antimicrobial activity of camel's milk against pathogenic strains of *Eschrichia coli* and *Listeria monocytogenes*, *Int. J. Dairy Technol.*, 57: 39–43.

Doledec, S., Chessel D., 1987, Rythmes saisonniers et composantes stationnelles en milieu aquatique I- Description d'un plan d'observations complet par projection de variables, *Acta Œcologica, Œcologia Generalis*, 8: 403–426.

Elagamy, E.I., Ruppanner R., Ismail A., Champagne, C.P., Assaf R., 1992, Antibacterial and antiviral activity of camel's milk protective proteins, *J. Dairy Res.*, 59: 169–175.

Elagamy, E.I., 2000, Effect of heat treatment on camel milk proteins with respect to antimicrobial factors: a comparison with cows' and buffalo milk proteins, *Food Chem.*, 68: 227–232.

Elamin, F.M., Wilcox, C.J., 1992, Milk composition of Marjheim camels, *J. Dairy Sci.*, 75: 3155–3157.

Farah, Z., 1993, Composition and characteristics of camel milk, *J. Dairy Res.*, 60: 603–626.

Farah, Z., Rüegg, M., 1991, The creaming properties and size distribution of fat globules in camel milk, *J. Dairy Sci.*, 74: 2901–2904.

Ihaka, R., Gentleman, R., 1996, A language for data analysis and graphics, *J. Comput. Graph.*, 5: 299–314.

Konuspayeva, G., Faye, B., 2004, A better knowledge of milk quality parameters: a preliminary step for improving the camel milk market opportunity in a transition economy – The case of Kazakhstan. Pages 28–36 in Proc. of Intl. Conf. "Saving the Camel and Peoples' Livelihoods". Sadri, Rajasthan, India 23–25 November 2004.

Konuspayeva, G., Loiseau, G., Faye, B., 2004, La plus-value "santé" du lait de chamelle cru et fermenté: l'expérience du Kazakhstan. Pages 47–50 in Proc. 11th Rencontre autour des Recherches sur les Ruminants, INRA, Paris 8–9 décembre, 2004.

Konuspayeva, G., Faye, B., Loiseau, G., Levieux, D., 2007, Lactoferrin and Immunoglobin content in camel milk from Kazakhstan, *J. Dairy Sci.*, 90: 38–46.

Konuspayeva, G., 2007, Variabilité physico-chimique et biochimique du lait des grands camélidés (Camelus bactrianus, Camelus dromedarius et hybrides) au Kazakhsta, Thèse en Sciences des aliments, Université de Montpellier II, France.

Sawaya, W.N., Khalil, J.K., Al-Shalhat, M., Al-Mohammed H., 1984, Chemical composition and nutritional quality of camel milk, *J. Food Sci.*, 49: 744–747.

Sharmanov, T.S., Dzhangabylov, A.K., 1991. Лечебные свойства кумыса и шубата (medical properties of kumis and shubat). Ed. Gylym, Almaty, Kazakhstan, 173 p.

Terenytev, C.M., 1975, Верблюдоводство (Camel farming). Ed. Kolos, Moscow, 224 p.

SELENIUM AND POULTRY PRODUCTS: NUTRITIONAL AND SAFETY IMPLICATIONS

DONATA CATTANEO, GUIDO INVERNIZZI, MARIELLA FERRONI, ALESSANDRO AGAZZI, RAFFAELLA REBUCCI, ANTONELLA BALDI, VITTORIO DELL'ORTO, GIOVANNI SAVOINI
Department of Veterinary Science and Technology for Food Safety (VSA), University of Milan, Via Trentacoste, 2, 20134 Milano, Italy

Abstract: Selenium (Se) is both an essential nutrient for humans and animals and a toxicant at excess levels in foods. Its content in animal products reflects that of the feeds consumed.

In order to meet Se nutritional requirements, it is a common animal production technology to supplement livestock and poultry diets with selenium. As a consequence, supplemented animals yield animal products (e.g. milk, meat, eggs) with higher Se concentrations. Because of concern about safety of human consumers, the quantities of selenium that can be supplemented to food-producing animals are strictly regulated. Nutritional and safety implications of selenium supplementation in animal production are discussed. Many studies indicate that selenium from organic sources, e.g. from selenium-enriched yeasts, is more bioavailable compared to inorganic sources and results in more selenium deposited in animal tissues. Problems can arise from accidental overdosing or errors in formulation of Se supplements in food-producing animals, which could lead to animal toxicosis and excessive enrichment of Se in the human food chain. A study is reported in which the effects of a tenfold overdose of Se, administered through Se-enriched yeast, have been examined on selenium status and on Se deposition in eggs and edible tissues in laying hens.

Keywords: Selenium overdosing, selenium-enriched yeast, laying hens, tissue selenium deposition, eggs

1. Introduction

Selenium (Se) is a controversial trace element, being both an essential nutrient and a potent toxicant, when present at excess levels in feeds and foods. For this reason it has been sometimes called "the essential poison" (Reilly, 2006). Selenium was recognised a toxic mineral long before its essentiality for mammals and birds was discovered in the late 1950s. Natural occurring selenium toxicity, or selenosis, has been described in the past centuries in farm animals grazing on seleniferous soils or fed crops grown on these soils (Combs, 2001).

1.1. SELENIUM IN ANIMAL PRODUCTS

The Se content in animal-derived foods reflects that of the feeds consumed by the animals. Normally occurring levels in meat, eggs, milk are in the range of 0.1–0.4 mg Se/kg, while much larger amounts (up to 1.5 mg Se/kg) can be found in organ meats, e.g. liver, kidney (British Nutrition Foundation, 2001). The principle chemical form of selenium is animal tissues is selenocysteine, incorporated specifically into selenoproteins/selenoenzymes, unlike plant tissues where selenomethionine (SeMet) predominates (Reilly, 2006). Selenium enters the human food chain through the soil, being removed from soils by plants and micro-organisms. Concentration of biologically available selenium in soils, water and crops vary significantly among different areas of the world, due to natural (geological, geographical) and anthropogenic factors, such as industrial activities (e.g. mining, combustion of coal and petroleum products, incineration of industrial and municipal wastes) and agricultural practices (e.g. addition of Se to fertilizers, Se dietary supplementation to farm animals) (Oldfield, 1998, 1999).

In many parts of the world, levels of biologically available Se in soils are inadequate and native content in grain and forages may be insufficient to meet animal nutritional requirements. Due to these strong geographical variations in Se content of crops and feeds, dietary supplementation of livestock and poultry rations with extra Se is now a standard agricultural practice in many countries, even in situations where deficiencies of the element do not exist. Because of concerns about possible health effects related to excess selenium entering the human food chain, the amount of Se that can be added to the feed are limited by government regulation in many countries and permitted levels vary with a range of 0.1–0.5 mg Se/kg in diets (Reilly, 2006).

The margin between Se adequacy and chronic toxicity levels in the diet is very narrow, and for selenium errors in feed mixing or in formulation of mineral supplements are not rare, with potential adverse effects to consumers of animal products. Both accidental and long-term overdose of Se result in intoxication of farm animals. In general, it is accepted that animal selenosis may appear when dietary Se content exceeds 5 mg/kg of feed (EFSA, 2006a). In avian species, selenosis reduces egg production and hatchability, with a great incidence of chick deformities, i.e. lack of eyes and beak, distorted wings and feet (Latshaw et al., 2004). In laying hens, 5 mg/kg of inorganic dietary selenium (sodium selenite) has been indicated to be borderline toxic, significantly decreasing hatchability of fertile eggs and increasing Se levels in eggs and edible tissues (Ort and Latshaw, 1978).

1.2. SELENIUM-ENRICHED YEAST

Traditionally, the primary forms of Se as feed additives have been inorganic, e.g. Sodium selenite (SS). In recent years, organic sources of Se, i.e. from Se-enriched yeast have received considerable attention (Schrauzer, 2000, 2001). Selenium-enriched yeast is produced by growing *Saccharomyces cerevisiae* in a selenium-rich nutrient medium, under conditions of sulfur limitations. This encourages the uptake of Se to form Se-analogues of organic compounds of sulfur, i.e. SeMet (Reilly, 2006). The majority of the commercial preparations of Se-yeasts contain mainly SeMet. Levels of SeMet have been reported to range, in different preparations, from 54% to 74% of total Se (Rayman, 2004).

In general, organic compounds, such as SeMet, are absorbed and retained in body tissues more efficiently compared to inorganic sources, but are not as efficient in maintaining selenium status (Fairweather-Tait, 1997). Unlike sodium selenite, which is a good substrate for the formation of selenoproteins, SeMet is not used directly for specific selenoprotein synthesis, but is incorporated non-specifically into body proteins, in place of methionine, and thus stored in body tissues as a reserve of Se (Schrauzer, 2003). As a consequence, meat, eggs and milk proteins may contain significantly more Se when animals are fed organic Se sources enriched with SeMet compared to inorganic Se sources.

Recently, two products consisting of inactivated selenized *Saccharomyces cerevisiae* yeast strains, providing Se in an organic form (mainly SeMet), have received a positive opinion from the European Food Safety Authority (EFSA 2006a, b) for use as feed additives in food-producing animals, at maximum permitted levels of 0.5 mg Se/kg of complete feed.

In the present paper, an overdose study is reported in which the effects of excessive dietary Se from organic sources (Se-enriched yeast), administered tenfold the maximum permitted level, have been examined on selenium status and Se deposition in eggs and other edible tissues and organs of laying hens.

2. Materials and Methods

2.1. ANIMALS AND TREATMENTS

For the purpose of the study, two trials were conducted to compare the effects of an inorganic Se source (SS = sodium selenite) with two different organic sources of Se (SY A = Se-enriched yeast, strain A, SY B = Se-enriched yeast, strain B), fed at two dietary levels (1X or 10X), on Se status and on accumulation and distribution of Se in edible tissues (eggs, muscle, organs) of laying hens:

– In the first trial (**study 1X**), dietary levels of Se from organic sources (SY A or SY B) were below the authorised maximum level (0.5 mg Se/kg of complete feed).
– In the second trial (**study 10X = Se overdose study**), dietary levels of Se from organic sources (SY A or SY B) were 10 times the maximum permitted level (5 mg Se/kg of complete feed).

Ninety-six Isa Brown Warren laying hens, with an egg rate of 86.3%, were fed the same basal diet with no Se supplementation (CP 18.3%, EE 6.03%, NDF 10.6%, Ca 3.34%, on as-fed basis, Table 1), for 5 weeks (pre-treatment period, from week 17 to 22 of age). At the end of the pre-treatment period, all hens were randomly distributed to six dietary treatments of 16 hens each and allotted in $34 \times 40 \times 45$ cm cages (two birds per cage), in order to have eight replicates per experimental groups. The dietary treatments were: (1) C (control): basal diet without Se supplementation; (2) SS (inorganic Se): basal diet plus 0.4 mg/kg of Se from sodium selenite; (3) SY A (1X): basal diet plus 0.4 mg/kg of Se from Se-enriched yeast strain A; (4) SY B (1X): basal diet plus 0.4 mg/kg of Se from Se-enriched yeast strain B; (5) SY A (10X): basal diet plus 5 mg/kg of Se from Se-enriched yeast strain A; and (6) SY B (10X): basal diet plus 5 mg/kg of Se from Se-enriched yeast strain B. Chemical composition and actual Se levels in each experimental diet are reported in Table 2. The experimental period lasted 8 weeks (from week 22 to 30 of age).

2.2. SAMPLE COLLECTION AND ANALYSIS

On days 0, 18, 36 and 56 of the experimental period, two eggs per replicate from each treatment were collected at random and stored at 4°C until analyzed for Se content.

At the end of the experiment, eight animals per group were slaughtered at random. Samples of serum, liver, kidney, breast muscle and skin were collected and stored at − 20°C until analyzed for Se content.

TABLE 1. Composition of the basal diet (as fed basis)

Ingredients	(%)
Corn	54.0
Soybean meal	29.0
Calcium carbonate	9.5
Soybean oil	4.5
Monocalcium phosphate	1.4
Vitamin and mineral mix	1.0
Salt	0.4
DL-Methionine	0.2

TABLE 2. Chemical composition of the experimental diets and actual Se levels (as fed basis)

	C	SS	SY A (1X)	SY B (1X)	SY A (10X)	SY B (10X)
Humidity	7.30	8.40	8.20	8.40	8.2	8.4
Crude protein (%)	18.63	18.56	18.32	18.04	18.2	20.1
Ether extract (%)	5.78	6.39	5.63	6.06	6.3	5.2
NDF (%)	10.27	10.68	10.51	10.11	9.6	10.6
Ash (%)	11.17	11.52	11.50	11.75	11.7	9.9
Leusine (%)	0.79	0.54	0.50	0.49	0.5	0.5
Ca (%)	3.41	3.09	3.22	3.16	3.7	2.8
P (%)	0.40	0.37	0.35	0.35	0.3	0.4
Se (mg/kg)	**0.11**	**0.46**	**0.47**	**0.49**	**6.1**	**5.7**

The selenium assay was performed by inductively coupled plasma atomic emission spectrometry (ICP-AES) for blood serum, liver and kidney (Machat et al., 2002) and by flow injection atomic spectrometry-atomic absorption spectrometry (FIAS-AAS) for muscle, skin and egg content. Determination of serum glutathione peroxidase (GPX) activity was performed by Glutathione Peroxidase Assay Kit, Cayman Chemical Company, Ann Arbor, MI, USA.

2.3. STATISTICAL ANALYSIS

The cage (two hens) represented the unit for the statistical analysis. The data relative to selenium deposition in eggs were analyzed by a mixed repeated procedure of SAS (SAS Institute, 1990). The data relative to blood parameters and Se deposition in tissues were tested by variance analysis according to the procedure General Linear Models (GLM) of SAS. Differences were considered as significant at $P < 0.05$.

3. Results and Discussion

3.1. ACTUAL SE LEVELS IN THE EXPERIMENTAL DIETS

The Se analyses of the experimental diets are reported in Table 2. Analysed values indicated that the basal diet contained 0.11 mg/kg of naturally occurring selenium. In the **1X study**, actual Se levels in the diets were 0.46, 0.47 and 0.49 mg/kg of Se, respectively for SS, SY A and SY B diets. These values were slightly higher than the calculated ones, although they remained below the maximum permitted levels of 0.5 mg Se/kg of complete feed.

In the Se overdose study (**10X study**), actual Se levels were 6.1 and 5.5 mg/kg of Se, respectively for SY A and SY B diets.

3.2. SELENIUM STATUS OF LAYING HENS

Results from **study 1X** show that Se supplementation from either Se sources (inorganic and organic), at dietary levels below the maximum permitted levels, significantly increased selenium status of laying hens, assessed by serum Se concentrations and measurement of plasma glutathione peroxidase (GPX) activity, as a functional bio-marker (Table 3).

TABLE 3. Selenium status of laying hens

	C	SS	SY A (1X)	SY B (1X)
Serum Se (mg/kg, wet weight basis)	0.15B	0.21b	0.33Aa	0.31Aa
GPX (U/ml)	0.375Bb	1.309A	1.120a	0.993a
	C	SS	SY A (10X)	SY B (10X)
Serum Se (mg/kg, wet weight basis)	0.2C	0.2C	0.6B	0.9A
GPX (U/ml)	0.375B	1.309A	1.384A	1.215A

a, b: P < 0.05; A, B: P < 0.01

The level of the specific selenoprotein GPX was higher in the blood of hens fed SS compared to hens fed SY A or SY B, confirming that SS may be more efficient than organic sources of Se in maintaining selenium status (Fairweather-Tait, 1997). Oxidised inorganic forms of Se, e.g. sodium selenite, are more efficiently metabolized to hydrogen selenide (H_2Se), which is incorporated specifically into selenoproteins, compared to SeMet, some of which is non-specifically incorporated into body proteins before entering the regular Se metabolism (Combs, 2001; Rayman, 2004).

Results from **study 10X** show that with a tenfold overdose of Se administered through Se-enriched yeast, Se levels in serum of laying hens increased up to 0.9 mg/kg. In farm animals, high blood Se levels up to 25 mg/l are typical to acute selenosis, whereas chronic Se overdose usually results in blood Se concentrations of 1–4 mg/l (EFSA, 2006a).

3.3. SELENIUM DEPOSITION IN EGGS

Egg Se concentrations of each dietary treatment are presented in Table 4. Initial egg Se contents were similar among the dietary treatments.

TABLE 4. Selenium deposition in eggs

	C	SS	SY A (1X)	SY B (1X)
Egg Se start (mg/kg, DM basis)	0.6	0.7	0.8	0.8
Egg Se mean (mg/kg, DM basis)	0.5C	0.9B	1.6A	1.4A
Egg Se mean (mg/kg, wet weight)	0.117C	0.208B	0.374A	0.326A
	C	SSC	SY A (10X)	SY B (10X)
Egg Se start (mg/kg, DM basis)	0.6	0.7	0.7	0.9
Egg Se mean (mg/kg, DM basis)	0.5C	0.9C	5.5B	6.6A
Egg Se mean (mg/kg, wet weight)	0.117C	0.208C	1.260B	1.564A

A, B: P < 0.01

Results from **study 1X** show that Se supplementation from both inorganic and organic sources at dietary levels below the maximum permitted levels significantly increased (P < 0.01) Se concentrations in eggs. The Se-enriched yeast diets yielded egg Se levels that were significantly higher (P < 0.01) than those from the SS diet, indicating a more efficient Se retention in eggs from hens supplemented with organic compared to inorganic sources of Se. These results are consistent with the results reported also by other authors (Payne et al., 2005; Utterback et al., 2005; Pan et al., 2007).

Results of the **10X study** show that with a tenfold Se overdose through Se-enriched yeast, Se concentration in eggs increased up to values as high as 1.5 mg/kg, on a wet weight basis, more then tenfold the Se levels in the control eggs. These are particularly high values, considering that normal Se levels in eggs vary, among different countries, in the range of 0.09–0.40 mg/kg, depending on the Se content in feeds (McNaughton and Marks, 2002; Reilly, 2006). Maximum residue limits (MRLs) in foods for widespread trace elements such as Se, naturally present in environment, vegetable and animal tissues, are not set by regulations in the EU. However, in other countries, selenium is included among the metal contaminants in foods for which guidelines levels have been established. An example is the Australia New Zealand Food *Standards Code* (ANZFA, 2001), which has set the following generally expected levels (GELs) for Se in foods, in order to maintain the lowest achievable levels: 2 mg/kg for edible offal (liver, kidney) and 0.2 mg/kg for muscle meat. In Canada, the maximum accepted level of selenium in tissues entering the human food chain is 2 mg/kg (Helie and Sauvageau, 1998).

Upper tolerable levels (ULs) for Se dietary intake by humans have been fixed by the EC Scientific Committee on Food (2000) at 300 µg Se/day for the adults. Consumption of one egg, typically weighing 36 g, from hens supplemented with tenfold Se overdose in our trial would result to a Se intake of 56 µg Se.

3.4. SELENIUM DEPOSITION IN EDIBLE TISSUES AND ORGANS OF LAYING HENS

Tissue Se concentrations of each dietary treatment are presented in Table 5.

TABLE 5. Selenium deposition in tissues and organs of laying hens (mg/kg, wet weight basis)

	C	SS	SY A (1X)	SY B (1X)
Breast muscle	0.115Bb	0.109Bb	0.346A	0.264a
Liver	0.409B	0.512B	0.603B	0.737A
Skin	0.09B	0.131b	0.163A	0.181Aa
Kidney	0.289	0.435	0.373	0.389
	C	SS	SY A (10X)	SY B (10X)
Breast muscle	0.115C	0.109C	1.087B	1.78A
Liver	0.409C	0.512C	2.002B	2.66A
Skin	0.09B	0.131B	0.50A	0.63A
Kidney	0.289C	0.435C	1.039B	1.485A

a, b: $P < 0.05$; A, B: $P < 0.01$

At dietary Se levels below the maximum permitted level (**Study 1X**), tissue Se concentrations were significantly higher in breast muscle, liver and skin of hens fed organic compared to inorganic sources of Se. In the kidney, on the contrary, selenium deposition did not differ significantly among dietary treatments, but a tendency was observed towards lower values in SY compared to SS fed hens. Similar results have been reported also by Pan et al. (2007). Kidney Se content reflects the amount of Se deposited in kidney and Se excreted through the urinary route (Pan et al., 2007). Higher Se excretion by kidney probably occurs when inorganic sources of Se are fed, whereas supplementation with organic Se sources determines a shift from excretion to retention of Se, as reflected by the higher Se content in tissues and organs (liver, muscle, eggs).

Results from **study 10X** show that with a tenfold overdose of Se administered through Se-enriched yeast, the concentration of selenium in edible tissues and organs of laying hens can reach high values, up to 1.5–1.8 mg/kg in kidney and breast muscle and 2.7 mg/kg in liver, from 5- to 15-fold the content in control tissues. Selenium levels reported for chicken meat from different countries normally vary in the range of 0.06–0.28 mg/kg (McNaughton and Marks, 2002; Reilly, 2006). The levels of tissue Se in our overdose trial are considerably higher than guidelines levels (2 mg/kg in edible offal, i.e. liver, kidney, and 0.2 mg/kg in muscle meat) mentioned before (ANZFA, 2001). Our results confirm that organ meats, such as liver, can accumulate significantly large amounts of Se.

Calculation of Se intake from edible tissues from over-supplemented hens in our trial, without considering background Se intake from other dietary sources, but only through the consumption of 105 g breast muscle, 35 g liver and 3.5 g kidney, corresponding to real consumption figures or SCOOP data (EC, 2004), can result in total loads to the consumers of 285 µg, near the UL of 300 µg Se/day set by EC SCF (2000) for the adults.

4. Conclusion

Selenium in animal-derived foods originates from inorganic and organic compounds present in feed sources. There are potential risks to humans from consuming products from animals that ingested excess levels of Se, either through accidental overdosing, or errors in formulation, or use in situations where Se deficiencies do not exist. The different metabolic fate of SeMet and sodium selenite, commonly used as Se supplements in animal nutrition, determines Se distribution and retention in animal tissues. Unlike sodium selenite, SeMet from Se yeast has the capacity to incorporate non-specifically into animal proteins and to accumulate very efficiently in animal products (meat, eggs).

Results from our Se overdose study indicate that Se at tenfold the maximum permitted level, supplemented through Se yeast, can increase Se deposition in edible tissues of laying hens at levels as high as 1.5 – 1.8 mg/kg for eggs, kidney and breast muscle and 2.7 mg/kg for liver. Chronic ingestion of large quantities of eggs and edible tissues (liver, kidney, meat) from hens that ingested excess Se could result in humans consuming more Se than is considered safe. Our study confirmed the ability of Se compounds from Se yeast to create significant Se deposits in animal tissues.

References

ANZFA, 2001, Australia New Zealand Food Authority, http://www.foodstandards.gov.au/thecode/foodstandardscode.cfm

British Nutrition Foundation, 2001, Selenium and health, Briefing Paper, British Nutrition Foundation, London, England, p. 5.

Combs, G., 2001, Selenium in global food systems, *Brit. J. Nutr.*, 85: 517–547.

EC Scientific Committee on Food, 2000, Opinion of the Scientific Committee on Food on the tolerable upper intake level of selenium, http://ec.europa.eu/food/fs/sc/scf/out80_en.html

EC European Commission, 2004, Reports on tasks for scientific cooperation (SCOOP) Task 3.2.11. Assessment of the dietary exposure to arsenic, cadmium, lead and mercury of the population of the EU Member States Directorate General health and Consumer Protection.

EFSA European Food Safety Authority, 2006a, Opinion of the scientific panel on additives and products or substances used in animal feed on the safety and efficacy of the product Sel-Plex®2000 as a feed additive according to regulation (EC) No. 1831/2003, http://www.efsa.europa.eu/EFSA/efsa_local-1178620753812_1178620782969.htm

EFSA European Food Safety Authority, 2006b, Opinion of the Panel on additives and products or substances used in animal feed on the safety and efficacy of the product Selenium enriched yeast (Saccharomyces cerevisiae NCYC R397) as a feed additive for all species in accordance with Regulation (EC) No. 1831/2003, http://www.efsa.europa.eu/EFSA/efsa_local-1178620753812_1178620781925.htm

Fairweather-Tait, S., 1997, Bioavailability of selenium, *Eur. J. Clin. Nutr.*, 51: S20–S23.

Helie, P., Sauvageau, R.A., 1998, Concern about the possibility of selenium in the human food chain, *Can. Vet. J.*, 39: 742.

Latshaw, J.D., Morishita, T.Y., Sarver, C.F., Thilsted, J., 2004, Selenium toxicity in breeding ring-necked pheasants (*Phasianus colchicus*), *Avian Dis.*, 48: 935–939.

Machat, J., Kanicky, V., Otruba, V., 2002, Determination of selenium in blood serum by inductively coupled plasma atomic emission spectrometry with pneumatic nebulization, *Anal. Bioanal. Chem.*, 372: 576–581.

McNaughton, S.A., Marks, G.C., 2002, Selenium content of Australian food: a review of literature values, *J. Food Comp. Anal.*, 15: 169–182.

Oldfield, J.E., 1998, Environmental implications of uses of selenium with animals, In: Environmental chemistry of selenium, W.T. Frankenberger Jr. and R.A. Engberg (Eds). Marcel Dekker, New York, pp. 129–141.

Oldfield, J.E., 1999, Selenium World Atlas, Selenium-Tellurium Development Association, Grimbergen, Belgium, p. 39.

Ort, J.F., Latshaw, J.D., 1978, The toxic level of sodium selenite in the diet of laying chickens, *J. Nutr.*, 108: 1114–1120.

Pan, C., Huang, K., Zhao, Y., Qin, S., Chen, F., Hu, Q., 2007, Effect of selenium source and level in hen's diet on tissue selenium deposition and egg selenium concentration, *J. Agric. Food Chem.*, 55: 1027–1032.

Payne, R.L., Lavergne, T.K., Southern, L.L., 2005, Effect of inorganic versus organic selenium on hen production and egg selenium concentration. *Poultry Sci.*, 84: 232–237.

Rayman, M.P., 2004, The use of high-selenium yeast to raise selenium status: how does it measure up? *Brit. J. Nutr.*, 92: 557–573.

Reilly, C., 2006, Selenium in food and health, 2nd ed. Springer, New York, p. 206.

SAS Institute, 1990. SAS User's guide: Statistics, Release 6.08. SAS Institute, Cary, NC.

Schrauzer, G.N., 2000, Selenomethionine: a review of its nutritional significance, metabolism and toxicity, *J. Nutr.*, 130: 1653–1656.

Schrauzer, G.N., 2001, Nutritional selenium supplements: product types, quality, safety, *J. Am. Coll. Nutr.*, 20: 1–4.

Schrauzer, G.N., 2003, The nutritional significance, metabolism and toxicology of selenomethionine, *Adv. Food Nutr. Res.*, 47: 73–112.

Utterback, P.L., Parsons, C.M., Yoon, I., Butler, J., 2005, Effect of supplementing selenium yeast in diets of laying hens on egg selenium content, *Poultry Sci.*, 84: 1900–1901.

EFFECT OF EXCESS SELENIUM ON DROMEDARY CAMEL IN THE UNITED ARAB EMIRATES

RABIHA SEBOUSSI, GHALEB AL-HADRAMI, MOSTAFA ASKAR
UAE university, Al-Ain, United Arab Emirates

BERNARD FAYE
Département environnement et société, CIRAD, Montpellier, France

Abstract: Early interest in selenium by nutritionists was first identified in the 1930 s as a toxic element, nowadays it is known to be important in livestock and human diet. Its poisonous nature arouses the curiosity of researchers to investigate the impact of this element in human and animal metabolism. However, selenium has become the center of attention due to its physiological functions explained on the basis of its role as an active component of the enzyme glutathione peroxidase (GSH-PX), which is responsible for the animal antioxidant defense by destruction of hydrogen peroxide and lipid peroxides. Selenium metabolism and toxicity has been consistently studied in different species but data investigations on camelidae species are very limited. Our current study is configured to investigate the selenium intolerance in dromedary camel and carry out the symptoms related to continuous selenium supplementation. Investigations showed that camel is potentially sensitive to selenium excess. Several symptoms revealed by their different intensity from 3 batches, resumed in alopecia – abnormal movement and posture, breathing difficulties, prostration, diarrhea, lost of weight and nervous alteration.

Keyword: Camelus dromedarius, selenium, toxicity, United Arab Emirates

1. Introduction

Essential role of selenium for animal has been well established by epidemiological studies. Nowadays, it's known that trace elements contribute to maintain animal health especially selenium, a key component of several enzymes such as glutathione peroxidase (GSH-Px) by inactivate free radicals (hydrogen peroxide) (Rotruck et al., 1973). Trace minerals are available from several sources- feedstuffs, drinking water and commercial supplement. This element tends to be localized mainly in the protein fraction of plant. In United Arab Emirates, camels' selenium intake comes mainly from diet, and selenium supplementation is often necessary by using different methods – injection, drenching and trace mineral salt mixes. While there is no established recommended daily allo-wance, this supplementation is done in anarchically way resulting by camel selenium overdosing which lead to a several sudden death in camel husbandry.

2. Experimental Procedure

This experiment was carried out at the United Arab Emirates, at the farm of nutrition and agriculture Faculty. Twelve healthy young 2-years camels shared into three batches were fed with similar basal diet including Rhodes grass (*Chloris gayana*) and commercial concentrate. They were supplemented by oral route for 3 months with selenium as sodium selenite respectively 8 mg (17.44 mg sodium selenite), 12 mg (26.16 mg sodium selenite) and 16 mg (34.88 mg sodium selenite). Selenium was given enrobed in date every day at the same time. The experiment began after 15 days of pretreatment. Body weight was taken every 2 weeks. Blood, was taken via jugular vein in heparinized and non heparinized blood collecting tubes after disinfection of local area. One animal of each batch was sacrificed at day 46 of selenium supplementation and another one at the end of the experiment at day 91. Different organs were taken to evaluate the selenium concentration and histopathology findings.

3. Results

Camels showed a characteristic signs of toxicity at 2 weeks of selenium supplementation in the three batches, starting with alopecia on the neck and enlarged to the abdomen, at the base of the tail and accompanied with hair discoloration (Photo 1).

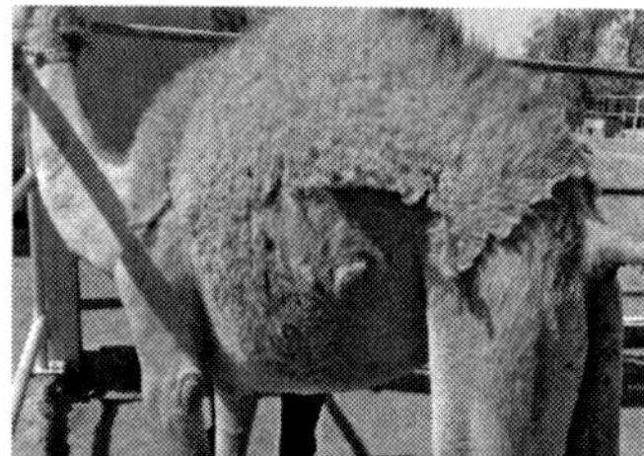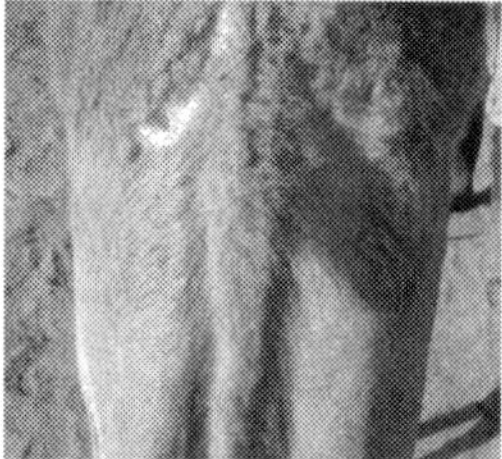

Photo 1. Alopecia on abdomen and at the base of the tail

Tears, hypertrophy of lymphoid gland less prominent in batch taking 8 mg and more remarkable in batches having 12 mg and 16 mg (Photo 2).

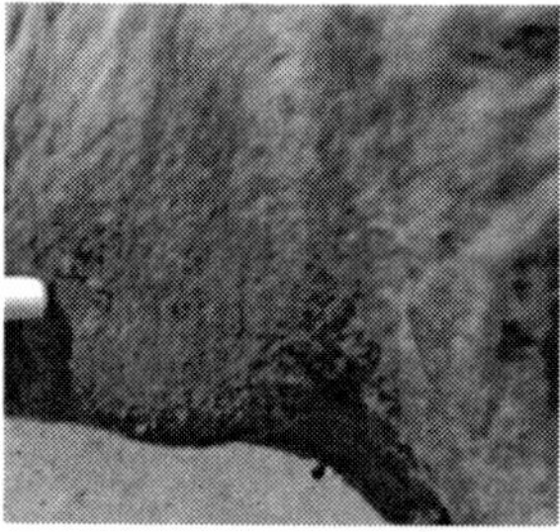

Photo 2. Hypertrophy of lymphoid gland in camel supplemented with 12 mg Se daily

After 45 days of supplementation till the end of experiment camels showed loss of appetite which lead to remarkable lost of weight, vesicular stomatitis (Photo 3), repeatable urination, diarrhea, nervous disorder with salivation and respiration become dyspneic.

Supplementation was stopped in batch 3 (16 mg) at day 45 to ovoid camel's death, and were treated with hepatoprotector. One camel of each batch was sacrificed to carry out the organs lesions and the histopathological finding. Camels shown hydrothorax,

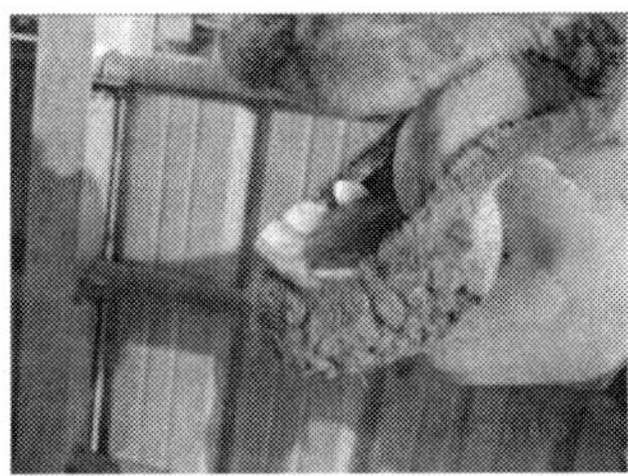

Photo 3. Vesicular stomatitis in Se intoxicated camel

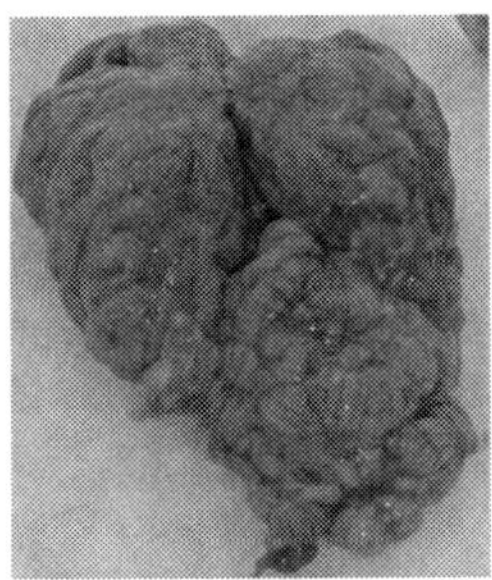 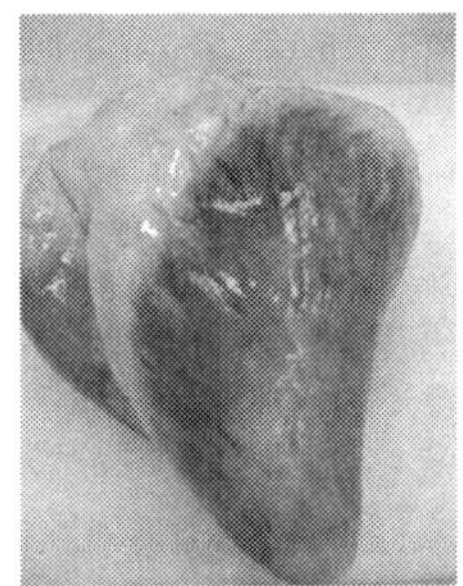 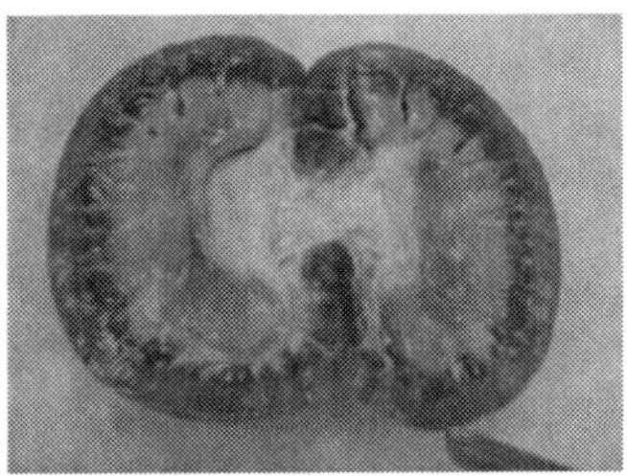

Photo 4. Brain edema, discoloration of heart and kidney in Se intoxicated animal

muscle discoloration and was more remarkable in diaphragm and intercostals' muscles, brain and lung edema, paleness of heart and kidney, enlargement of spleen (Photo 4). Congestion and necrosis of lung, heart, kidney, spleen and liver was observed.

The mean plasma values in intoxicated animals were respectively 321 ± 140 ng/ml (8 mg Se supplementation), 443 ± 231 ng/ml (12 mg) and 298 ± 212 ng/ml (16 mg). The extremes values were between 16 ng/ml at the beginning of the experiment up to 890 ng/ml that was ninefold the normal value.

4. Discussion

Selenium toxicity was revealed in all livestock but to our knowledge, no reported publication in camel selenosis. There are two forms of selenium toxicity: acute and chronic. The acute toxicity is due to ingestion of seleniferous plants, which provokes death in few hours. The chronic toxicity is responsible of "Blind staggers" and "alkali disease" (McDowell, 1989). It is not limited to grazing livestock but can occur from consumption of supplemented diet which contain high selenium amount. If selenium intake is prolonged, camels show clinical symptoms as indicated in our study. It seems almost identical in ruminants. Selenium is rapidly distributed to major organs of the camel. We consider that central nervous system is the first target organ, also liver, heart, lung are affected. Dyspnea symptoms observed in camels were revealed in sheep injected with selenium by Blodget (1987), and the lesions were edematous lungs and pale mottled, degenerated and necrosis hearts. It was reported by Mihajlovic (1992) that clinical signs of chronic selenosis in horse, cattle and swine are loss of hair (horses and cattle lose long hair from the mane and tails), emaciation, hoof lesions and lameness. Sheep showed lose of appetite and have reduced weight gain. Skin lesions and alopecia, hoof necrosis have also been reported by Harr and Muth (1972). These symptoms reported in the literature confirm our results and indicate that prolonged intake of selenium as sodium selenite by camel can provoke a chronic selenosis. These clinical symptoms are linked to a high level of selenium in plasma compared to values on non supplemented animals that were between 129 and 282 ng/ml according to their physiological status (Seboussi et al., 2004).

5. Conclusion

Our study revealed that camel was potentially sensitive to prolonged selenium intake and as consequence it could develop a chronic selenosis and shared most of toxicity symptoms with other ruminants reported by researchers. It has been shown that the amount of selenium needed to cause chronic toxicity in camels was 8 mg (17.44 mg sodium selenite). We concluded that duration of supplementation and concentration of the element are conditional factors of camel response to the selenium supplementation. Physiological status of the camel, stress and environmental factors could affect the trace elements metabolism. There was a need to develop the toxicity metabolism in camel species and to validate the factors that could affect camel response to selenium supplementation.

Acknowledgments

This study was carried out under the frame of cooperation between France and the United Arab Emirates, upon the agreement signed between CIRAD and the United Arab Emirates University. We thank the international cooperation bureau at the French embassy Abudhabi. Our sincerest acknowledgments to the Agriculture and municipalities department, Al-Ain for their help and facilitating the use of Swehan slaughter house.

References

Blodget, D.J, 1987, Acute selenium toxicosis in sheep, Chemicals Metabolism and Toxicology – Vertebrates, Report, Wisconsin, USA.

Harr, J.R., Muth, O.H., 1972, Selenium poisoning in domestic animals and its relationship to man., *Clin Toxicol.*, 5(2): 175–186.

McDowell, L.R., 1989, Vitamins in animal nutrition. Academic/Harcourt Brace Jovanovich, San Diego, CA.

Mihajlovic, M., 1992, Selenium toxicity in domestic animals Glas, Srpska akademija nauka i umetnosti, *Odeljenje medicinskih nauka*, 42: 131–144.

Rotruck, J.T., Pope, A.L., Ganther, H.E., Swanson, A.B., Hafeman, D.G., Hoekstra, W.G., 1973, Selenium: biochemical role as a component of glutathione peroxidase, *Science*, 179: 588–590.

Seboussi, R., Faye, B., Alhadrami, G., 2004, Facteurs de variation de quelques éléments trace (sélénium, cuivre, zinc) et d'enzymes témoins de la souffrance musculaire (CPK, ALT et AST) dans le sérum du dromadaire (Camelus dromedarius) aux Emirats Arabes Unis, *Rev. Elev. Med. Vét. Pays Trop.*, 57(1–2): 87–94.

CONTAMINATION BY PERSISTENT CHEMICAL PESTICIDES IN LIVESTOCK PRODUCTION SYSTEMS

BRUNO RONCHI AND PIER PAOLO DANIELI
DIPAN, Università della Tuscia, Viterbo, Italy

Abstract: The impact of persistent organic pesticides represents one of the major environmental problems as reported in several studies and reflected in some mandatory actions at the inter-governmental level. In particular, isomers of Hexachlorocyclohexane (HCH), like many others Organochlorine Pesticides (OCPs), are of human health and environmental concern due to their persistence in the biosphere. In industrialised countries, Lindane (the γ-isomer of HCH with insecticide effects) has been widely used in the past. As a consequence, large quantities of HCH isomers without insecticide effects (α-, β-, δ- and ε), discarded during the purification of Lindane, have been disposed for years into dumps around industrial sites. Nowadays, such uncontrolled disposal practices imply a great risk of environmental contamination, possibly threatening animals and humans by food chains transfer and bio-accumulation. In 2005, dairy cow farms in the Province of Rome (Italy) were threatened due to environmental contamination of the Sacco River by HCHs. As a case study, all components of the dairy cow production system undergoing agro-environmental pollution crisis, were investigated with the aim to analyze the main critical points. Five dairy farms were involved in the research. Data regarding the contamination by HCHs of soils, forages, bovine milk and blood serum are reported. Soil and forage samples (mainly maize, alfalfa and ryegrass) were taken in different places near the Sacco River, on the basis on irrigation practiced and flooding conditions. All samples were analyzed by Gas Cromatography using an Electron Capture Detector. Soil contamination by HCHs was found higher nearby than away the river (p < 0.01) with a great incidence of outflow risk (p < 0.01), while no differences were observed on the basis of irrigation practices. In alfalfa samples higher concentration of HCHs than in ryegrass were detected, with a greater plant/soil apparent partition factor. Differences in milk contamination by β-HCH among dairy farms (p < 0.01) and sampling time were found (p < 0.05). In many cases, the β-HCH content of milk resulted above the EU limit (0.003 mgβ-HCH/kg), posing serious hazard for human consumption due to chronic toxicity of that isomer. Differences in milk β-HCH concentration were found related to lactating phase and parity. A linear regression between blood serum and milk β-HCH concentration was observed ($r^2 = 0.919$, p < 0.05). Furthermore, β-HCH as a trace was detected in blood sera when milk levels fell below the analytical limits, indicating the usefulness of blood serum HCH content as an early indicator of animal exposure. Results obtained from the case study highlighted the needing for further researches at wider level, due to the strong impact of persistent organic pesticides on rural environment and human food chain.

Keywords: POPs, HCHs, dairy farming, agro-environmental pollution

1. Introduction

Persistent organic pollutants (POPs) are mainly anthropogenic in origin, heavily used and released into the environment in large amount (Breivika et al., 1999; Misra et al., 2007), often at rates not sustainable by natural degradation processes. As a consequence, they show long natural half-lives with a progressive increase of their global storage in the environment. POPs can be transported through ecosystems and travel distances on a local-to-global scale mainly trough the air (Tanabe et al., 1982, 1983; Simonic and Hites, 1995; Wania and Mackay, 1996; Harrad and Mao, 2004), and secondarily trough water as transport medium (US-EPA, 1994; Nikanorov, 2007). Extensive scientific studies have shown that POPs may be regarded as of the most hazardous substances released into the environment by human (Tanabe et al., 1994; Word Bank, 2001; Thomas et al., 2005). Among POPs, there is a real concern over the residual properties of the organochlorine pesticides (OCPs) class, widely used as insecticides for livestock and crop protection (Jolly, 1967; Spiric and Saicic, 1998). Such compounds are able to accumulate especially in the animal fatty tissues (Carter et al., 1953), showing a tendency to concentrate along food chains (Clark et al., 1988; Harding et al., 1997). Lindane, the commercial name of the γ-isomer of the OCP hexachlorocyclohexane

(HCH), was produced since the 1950s and sold mainly as seed dressing pesticide in industrialized countries (Garmouna and Poissant, 2004). The insecticide property of γ-HCH is not shared with other stable HCH isomers, namely the α-, β-, δ- and ε-HCH, synthesized in variable percentage as by-products of the industrial manufacture of Lindane (Hayes, 1982; UNEP, 1995). Otherwise, these isomers show similar toxicity of Lindane for mammals (Metcalf, 1955). A less expensive commercial formulation including all HCH isomers, named "technical HCH", was widely sold and used in developing countries, mainly in Asia, for crop protection, sanitation and vector born disease control (Iwata et al., 1993; John et al., 2001). World technical HCH production between 1945 and 1992 was estimated to be 1,400,000 t (Barrie et al., 1992). In 1976, United States banned technical HCH for agricultural purposes (ATSDR, 2003). Many European countries, severely restricted or banned technical HCH and Lindane between the years 1977 and 1992 (Breivika et al., 1999). These authors estimated a cumulative European usage of technical HCH and Lindane from 1970 to 1996 of 382,000 and 81,000 t, respectively. Taking into account that γ-isomer represents a relatively small fraction (8–15%) of the chlorination products of benzene (Metcalf, 1955; UNEP, 1995), a great amount of waste containing mostly the α-, β- and δ- isomers have been addressed (in licit and illicit ways) toward the disposal (Nerin et al., 1991; Loibner et al.,[1] 1998; Frazar, 2000; Langenhoff et al., 2002; Wenzel et al., 2006). Nowadays, a large number of contaminated sites exist in Europe, estimated at 250,000 by EAA (2007).[2] A relevant portion of these sites, roughly the 2.4% (over 6,000 sites), could represent a potential hazard for ecosystems as far as the OCPs contamination is concerned (EAA, 2007). Moreover, in European Union some industrial activities (basic organic chemistry, biocides/explosives industry, paper manufacture and installation for disposal of hazardous/urban wastes) still represent the main official emissive sources of HCHs in water and air, with an estimated yearly release of 0.029 and 0.25 t respectively (EC-EPER, 2004).[3] To date, a massive literature dealing with the presence of OCPs in human, rural and natural environments is available: Romania (Covaci et al., 2001), Serbia (Skrbic' and Durisic-Mladenovic, 2007), the UK (Meijer et al., 2001), Germany (Manz et al., 2001), Spain (Losada et al., 1996), Russia (Nikanorov et al., 2007), Italy (Galassi et al., 1996; Naso et al., 2004), Turkey (Ayas et al., 1997), the Netherlands (Langenhoff et al., 2002), Greece (Mallatou et al., 2002), Ireland (McGrath, 1995), India (Hans et al., 1999; Sharma et al., 2007), Corea (Yeo et al., 2004), Mexico (Waliszewski et al., 1997), USA (Shen et al., 2005), Australia (Van Barneveld, 1999). In spite to the relevance of the food chain accumulation and transfer to humans of OCPs, and the figures of potentially contaminated sites or anthropogenic activities leading to a release of these compounds, no detailed appraisals on episodes dealing with livestock production systems threatened by OCPs were found in the literature. Objective of the present work is to analyze a case study of agro-environmental contamination by HCHs in Italy entailing livestock systems at the beginning of 2005. During that episode, a large rural area devoted to dairy cow farming in Central Italy, was discovered to be contaminated by HCH isomers. In the present paper are reported the results of a study

---

[1] Loibner A.P., Farthofer M., Braun R., 1998. *Aerobic degradation of hexachlorocyclohexane-isomeres in soil monitored by using an on-line GC/MS system.* In: Proc. 4th Int. Symp. Exhib. Environ. Contam. Cent. East Eur., Warsaw, 548–552.

[2] EAA, 2007. *Progress in management of contaminated sites (CSI 015)* – Assessment published August 2007. The European Environmental Agency, Copenhagen K, Denmark. http://themes.eea.europa.eu/IMS/ISpecs/ISpecification20041007131746/IAssessment1152619898983/view_content

[3] EC-EPER, 2004. *Final Report.* European Commission Dg Environment. http://www.eper.ec.europa.eu/eper/

investigating the presence of HCHs in soils, bovine milk and blood samples obtained from dairy farms located in the Sacco River Valley (Province of Rome, Lazio Region).

2. Materials and Methods

2.1. SAMPLING

Five dairy cow farms were selected (Figure 1) in the polluted floodplain of the Sacco River, on the basis of a previously official survey of milk contamination by β-HCH (IZSLT, 2005).[4] In each farm, from July 2005 to September 2005, every 3 weeks milk and blood samples were collected from lactating cows at different lactation phase (first: from 10 to 100; second: from 101 to 200; and third lactation: last 100 days) and parity (from first to sixth lactation). From March 2005 dairy cows were fed on forages and concentrates without detectable contamination by HCHs.

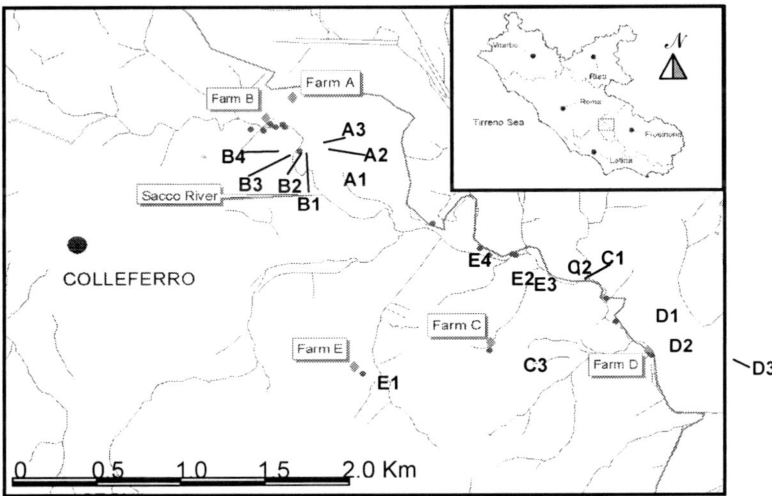

Figure 1. Position of the dairy farms investigated in the Province of Rome (green solid line). Labelled round dots indicate the sites where soils and vegetables were sampled for the assessment of agro-environmental contamination by HCHs

Individual milk samples were taken during the morning milking using calibrated weigh jars. Milk samples were stored at −20°C until analysis. Venous blood samples were obtained using vacutainer tubes without anticoagulant agent (Becton, Dickinson & Company, Plymouth, UK). The samples were left naturally coagulate at 4°C and then centrifuged to separate the serum which was stored at −20°C until analysis.

On selected sites (judged representative of farm polluted/unpolluted soils) three randomly logs were made within a 10 m square sampling patch by means of a hand soil logger. For each farm site were recorded data about overflowing frequency, irrigation practices and type of crops cultivated in the 3 years period before sampling time. Logs were performed separating the 0–5 cm depth, 5–20 cm depth and 20–30 cm depth soil

---

[4] IZSLT, 2005. *Contaminazione delle produzioni zootecniche da beta-esaclorocicloesano nel territorio delle produzioni zootecniche da beta-esaclorocicloesano nel territorio del bacino del fiume sacco.* http://www.izslt.it/izs/modules/sections/index.php?op=viewarticle&artid=61

layers as different samples. Within a patch, the three replicates for each soil layer were pooled and sub-sampled for analysis. From each site, forage samples were collected for quantify the HCH accumulation.

2.2. LABORATORY ANALYSES

2.2.1. Samples treatment and clean-up

Pre-treatment of vegetables, feedstuffs and soils were carried out adapting the method described by Hans et al. (1999). Representative 100 g dry sample aliquots were grounded and extracted twice by shaking in 150 ml HPLC grade *n*-hexane and *iso*-propanol mixture (2:1, v/v) (Sigma, Germany) for 30 min. Filtered extracts were washed with saturated NaCl solution in water (30 ml/100 ml extract) and the hexane layer was treated with 2 ml of concentrated sulphuric acid (Signa-Aldrich, Germany) allowing to stands for 15 min with occasional shaking. The hexane layer was separated and traces of sulphuric acid were removed by washing twice with water (equal volume). Clean-up was performed passing washed hexane extract through an activated Forisil© column (Fluka, AG, Buchs, Switzerland). After drying under a pure nitrogen stream (Rivoira, Italy), extracts were recovered with 500 µl of *iso*-octane (Merk, Darmstad, Germany). Milk samples (20 ml) were extracted with 40 ml acetone-petroleum ether mixture (Sigma-Aldrich, Germany) by shaking for 1 h. After a preliminary volume reduction of the extract under nitrogen, a clean-up step was performed using Florisil© as adsorbent matrix and the purified extract concentred to 2 ml under a nitrogen for submission to GC-ECD analysis. Blood serum treatment was performed as reported by Otero et al. (1997). Briefly, 2 ml of blood serum were digested adding 3 ml of *n*-hexane (Sigma-Aldrich, Germany) and 2 ml of concentrated sulphuric acid (Sigma-Aldrich, Germany) and then stirred for 30 s. The mixture was cooled at room temperature adding five drops of acetone to help phase separation. The supernatant *n*-hexane was collected in a polyethylene 10 ml tube. The remaining sulphuric solution was re-extracted twice with 2 ml of *n*-hexane, under the same conditions. The final 7 ml *n*-hexane phase collected has been washed with 2 ml of sulphuric acid and then concentrated to almost dryness under nitrogen at room temperature. Prior to be submitted to GC-ECD analysis, cleaned-up samples were recovered with *iso*-octane to a 500 µl final volume.

2.2.2. HCH determination by GC-ECD

Pre-treated samples were analyzed using a Hewlett-Packard 5890A model gas-chromatograph equipped with a 30 m × 0.53 mm I.D., 1.5 µm film thickness DB1 capillary column (J&W Scientific, Folsom, CA, USA) and an ECD system for detection of HCH isomers. A second 30 m × 0.53 mm 1.0 µm film thickness DB17 capillary column (J&W Scientific, Folsom, CA, USA) was used as a confirmative column. The GC oven temperature was programmed to heats columns at 140°C (holding time: 2 min) and then rises to 180°C at 6°C/min, 195°C at 0.8°C/min and 280°C at 30°C/min, keeping the last temperature for 10 min. The injector and detector temperatures were set respectively at 280°C and 300°C. Injection (1.5 µl) was operated in split/splitless mode using helium (Rivoira, Italy) as carrier gas (20 ml/min) and injector make up gas (10 ml/min). Qualitative and quantitative analyses were made by comparing the retention time and peak area of the samples with those obtained for the calibration standards (Restek, Bellefonte, PA, USA). Trials with fortified samples gave mean recovery for HCH (sum of all isomers) of 80%, 88% and 78% respectively for solid samples, milk and blood serum. Detection limits (LOD) were estimated (DIN, 1994) as low as 0.002 mg β-HCH for soil and vegetables, 0.0002 mg β-HCH/kg for milk and 0.0002 μg β-HCH/ml in the case of blood serum. Quantification limits (LOQ) were estimated as 3.1 times the respective LOD values (DIN, 1994).

2.3. CLIMATIC DATA

During the study, daily temperature (min-max) and rainfall data were obtained from the A.R.S.I.A.L. Meteorological Station located at Anagni (352 m high), less than 7 km away from the study area. Hereafter, data were expressed as mean (temperature) and cumulative (rain) values per decade.

2.4. STATISTICAL ANALYSIS

Data were analyzed by ANOVA using a nested design to explore causal factors (sampling date, farm, parity and lactating phase) driving milk contamination. A non-nested ANOVA, instead, was used for the evaluation soil/plants HCH levels variability including as factors, the sampling depth, the dairy farm, the linear distance from the river (<50 m, 50–100 m and >100 m), the flooding risk (high, medium and low) and the irrigation practices (irrigated *vs.* un-irrigated). Linear Regression was performed to study milk *vs.* blood contamination by β-HCH. Effects were taken into account as significant at p-level below 5% (p < 0.05). To verify the normality of data distribution, a preliminary evaluation was performed applying the Shapiro & Wilks Test (Shapiro and Francia, 1975). Data treatment (square rot or log-transformation) was used as needed. All data analysis, were performed using the package STATISTICA™ Release 6.0 (StatSoft Inc., Tulsa, OK, USA).

3. Results and Discussion

3.1. SOIL AND FORAGES CONTAMINATION BY HCHS

In soil samples 5 isomers of HCH were found: α-, β-, γ-, δ- and ϵ-HCH. The α- and β-isomers resulted the most commonly HCH contaminants in soil samples (Table 1).

Considering the Italian limit (0.01 mg/kg) prescribed for HCH in soils intended for residential purpose, a quote of 39.2% and 21.5% of soil samples were exceeded the law thresholds for β- and α-HCH respectively. A clear pollution level *vs.* sampling depth relationship was not observed (p > 0.05). Positive soils samples resulted highly variable in HCH concentration. Soil samples collected in farms C, D and E were the most contaminated (Figure 2a). A rough contamination by HCHs *vs.* downstream site location relationship was found (Figure 2b), indicating the relative position along the river course as a probable explicative factor for differences in HCHs contamination of soils. A more strong relationship was observed for β-/α-HCH ratio on soils and the relative downstream site location. An interesting concordance of contamination data between soils and river sediments was ascertained on the basis of an official survey of water and sediments carried out by

TABLE 1. Contents of HCHs (mg/kg) in dry soil samples (sites located as in Figure 1). LOD: limit of detection. LOD for all isomers: 0.002 mg/kg DW

Site code	Depth (cm)	α-HCH	β-HCH	γ-HCH	δ-HCH	ε-HCH	ΣHCH
A1	0–5	0.0240	0.1150	<LOD	0.0058	0.0035	0.1483
	5–20	0.0030	0.0622	<LOD	<LOD	<LOD	0.0652
	20–30	0.0057	0.0960	<LOD	<LOD	0.0038	0.1055
A2	0–5	0.0070	0.0860	<LOD	<LOD	<LOD	0.0930
	5–20	0.0240	0.1231	<LOD	0.0068	<LOD	0.1539
	20–30	0.0840	0.2280	0.0089	0.0950	0.0310	0.4469
A3	0–5	<LOD	<LOD	<LOD	<LOD	<LOD	–
	5–20	<LOD	<LOD	<LOD	<LOD	<LOD	–
	20–30	<LOD	<LOD	<LOD	<LOD	<LOD	–
B1	0–5	<LOD	<LOD	<LOD	<LOD	<LOD	–
	5–20	0.0039	0.0096	<LOD	<LOD	<LOD	0.0135
	20–30	<LOD	<LOD	<LOD	<LOD	<LOD	–
B2	0–5	<LOD	<LOD	<LOD	<LOD	<LOD	–
	5–20	<LOD	<LOD	<LOD	<LOD	<LOD	–
	20–30	<LOD	<LOD	<LOD	<LOD	<LOD	–
B3	0–5	<LOD	0.0170	<LOD	<LOD	<LOD	0.0170
	5–20	<LOD	<LOD	<LOD	<LOD	<LOD	–
	20–30	<LOD	0.0033	<LOD	<LOD	<LOD	0.0033
B4	0–5	<LOD	<LOD	<LOD	<LOD	<LOD	–
	5–20	<LOD	<LOD	<LOD	<LOD	<LOD	–
	20–30	<LOD	<LOD	<LOD	<LOD	<LOD	–
C1	0–5	0.0064	0.1070	0.0024	<LOD	<LOD	0.1158
	5–20	0.0260	0.2912	0.0140	0.0047	<LOD	0.3359
	20–30	0.0410	0.3220	0.0067	0.0055	<LOD	0.3752
C2	0–5	0.0025	0.0513	0.0025	<LOD	<LOD	0.0563
	5–20	0.0028	0.0361	<LOD	<LOD	<LOD	0.0389
	20–30	<LOD	0.0210	<LOD	<LOD	<LOD	0.0210
C3	0–5	<LOD	<LOD	<LOD	<LOD	<LOD	–
	5–20	<LOD	<LOD	<LOD	<LOD	<LOD	–
	20–30	<LOD	<LOD	<LOD	<LOD	<LOD	–
E1	0–5	<LOD	<LOD	<LOD	<LOD	<LOD	–
	5–20	<LOD	<LOD	<LOD	<LOD	<LOD	–
	20–30	<LOD	<LOD	<LOD	<LOD	<LOD	–
E2	0–5	<LOD	<LOD	<LOD	<LOD	<LOD	–
	5–20	<LOD	<LOD	<LOD	<LOD	<LOD	–
	20–30	<LOD	<LOD	<LOD	<LOD	<LOD	–
E3	0–5	0.0410	0.5520	0.0042	0.0057	<LOD	0.6029
	5–20	0.0160	0.3250	<LOD	0.0035	<LOD	0.3445
	20–30	0.0170	0.3220	<LOD	0.0035	<LOD	0.3425
E4	0–5	<LOD	<LOD	<LOD	<LOD	<LOD	–
	5–20	<LOD	<LOD	<LOD	<LOD	<LOD	–
	20–30	<LOD	<LOD	<LOD	<LOD	<LOD	–
D1	0–5	0.0024	0.0350	<LOD	<LOD	<LOD	0.0374
	5–20	0.0033	0.0200	<LOD	<LOD	<LOD	0.0233
	20–30	<LOD	0.0045	<LOD	<LOD	<LOD	0.0045
D2	0–5	0.0221	0.4762	0.0058	<LOD	0.0037	0.5078
	5–20	0.0170	0.2850	0.0051	<LOD	0.0035	0.3106
	20–30	0.0220	0.2771	<LOD	0.0091	0.0067	0.3149
D3	0–5	<LOD	<LOD	<LOD	<LOD	<LOD	–
	5–20	<LOD	<LOD	<LOD	<LOD	<LOD	–
	20–30	<LOD	<LOD	<LOD	<LOD	<LOD	–

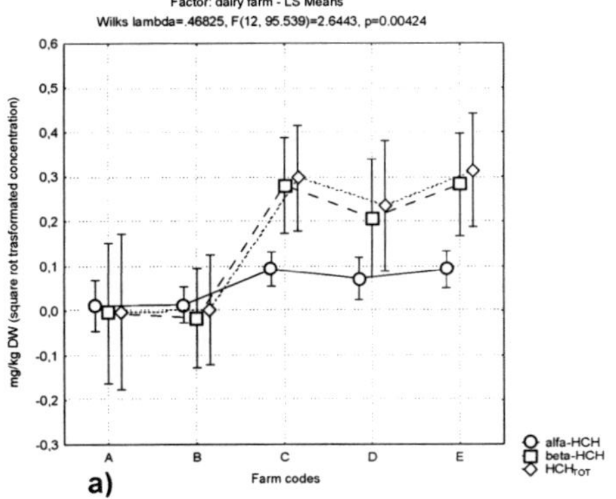

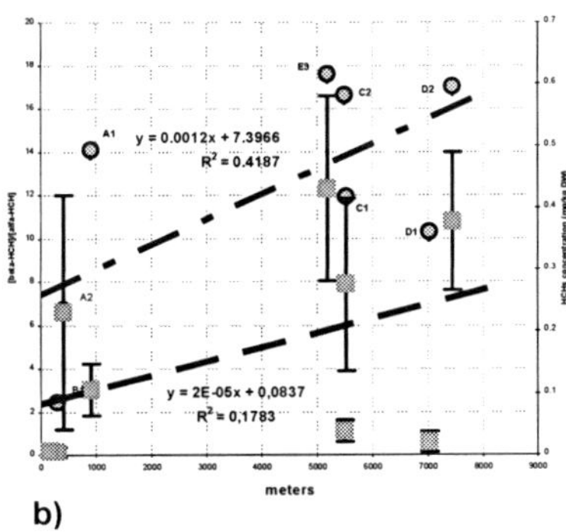

Figure 2. (a) LS Means of HCH concentration of soil samples (square rot transformed). Effect of the farm where samples have been collected. Vertical bars denote 0.95 confidence intervals. (b) Relationship between HCH concentration (squared dots) or β-HCH/α-HCH ratio (round dots) and relative distance of sites along the river course from the upstream one (Site B4)

the Regional Agency of Environmental Protection (data unpublished). From April to May 2005, 0.041 mg HCH/kg DW were found in sediments in a river section a bit upstream respect the location of soils sampled in the Farm B, while in two places located 8 km downstream (near soils collected for farms D and E), sediments resulted more heavily contaminated ranging from 0.118 mg HCH/kg DW to 0.191 mg HCH/kg DW. The downstream river sections resulted morphologically different from the upstream one, widely influenced by tributaries and by modifications made by rural population. As a consequence, it was highlighted that the flooding frequency (flooding risk) affects in significant manner the HCH presence in soil (Figure 3a). A direct relationship between distance from riverside and HCH contamination of soils may was hypothesized (Figure 3b) with an increase of contamination risk in the 50 m strip nearest the riverside. No significant effect on HCH contamination of soil ($p >$ 0.05) was found as far as the watering practice is concerned. Such results and observations, suggest that overflowing events followed by sediment deposition could be the main causal factor of soil pollution by HCHs. This hypothesis is in agreement with data reported by Nerin et al. (1991) about the HCH isomers distribution in soil, sediments and water along the Gallego River (Spain) released from a Lindane factory with a disposal facility. These researchers came to some interesting conclusions about the variability of distribution found particularly for α- and β-HCH. Among the environmental matrices investigated, the β-isomer was found strictly linked to river sediments. Despite the attended high level of the α-HCH contamination on the basis of pollution source, these authors observed a moderate concentration of that isomer in all samples analyzed, postulating a fast transformation into the others isomers, under specific environmental conditions. At the end, these authors highlighted the relevance of man-made modification of river course (dams), explaining spatially different HCH levels.

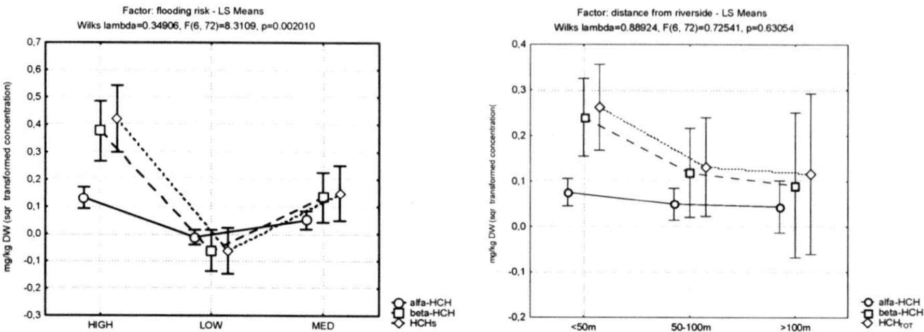

Figure 3. LS Means of HCH concentration of soil samples (square rot transformed). Vertical bars denote 0.95 confidence intervals: (a) effect of the flooding risk, (b) effect of the distance from the riverside of sampling sites

For some sites, vegetables collected and analyzed during our study, gave the HCHs contents reported in Table 2. Sometime a very high contamination of plant tissues (stems and leafs) was observed up to 1.6 mg β-HCH/kg DW found in an alfalfa sample (*Medicago sativa*, Fam. Leguminosae). No detectable amounts of HCHs were found in maize samples. The α-isomer was found only in one ryegrass sample (*Lolium multiflorum*, Fam. Gramineae) whereas in any case was recorded the presence of other HCH isomers. Soil content of HCHs did not appear influenced by forage crops. Between ryegrass and alfalfa cultivated soils, no significant differences were found ($p > 0.05$) (Figure 4a). Despite this fact, a significant difference was observed in the accumulation capacity of alfalfa and ryegrass respect the HCHs content of soils. If partition plant/soil coefficient is calculated as HCH level in plant tissues on level found in respective soil samples (as a mean value for the site), alfalfa showed the highest accumulative capacity (Figure 4b). Similar findings were reported by Concha-Grana et al. (2006) about the presence of chlorinated pesticides (HCHs and DDT) in tissues of plants grew on contaminate sites. A comparison of soil contamination and HCH levels in vegetables highlighted that, in some cases, plant roots seemed to concentrate HCHs and an interesting variability in plant species capability to accumulate HCH isomers was found. Among five different species, belonging to four botanical families, leguminous like *Cytisus striatus* and *Vicia sativa*, shown a higher accumulative capacity towards HCHs than that of other species belonging to Graminae, Solanaceae and Chenopodiaceae families. Moreover, experimental trials with artificially contaminated sites described by Verna and Pillai (1991), gave similar results with leguminous plants (chickpea) accumulating up to ten times more HCHs than grasses species (i.e.: maize and rice).

TABLE 2. Content of HCH isomers (mg/kg) in dry plant tissues (sites located as in Figure 1). LOD: limit of detection. LOD for all isomers: 0.002 mg/kg DW

Site code	Forage crop type	α-HCH	β-HCH	γ-HCH	δ-HCH	ε-HCH	ΣHCH
A1	*Medicago sativa*	<LOD	1.450	<LOD	<LOD	<LOD	1.450
A2	*Zaea mays*	<LOD	<LOD	<LOD	<LOD	<LOD	–
B1	*Lolium multiflorum*	<LOD	0.028	<LOD	<LOD	<LOD	0.028
B2	*Lolium multiflorum*	<LOD	<LOD	<LOD	<LOD	<LOD	–
B3	*Medicago sativa*	<LOD	0.063	<LOD	<LOD	<LOD	0.063
B4	*Lolium multiflorum*	<LOD	<LOD	<LOD	<LOD	<LOD	–
C1	*Lolium multiflorum*	<LOD	0.423	<LOD	<LOD	<LOD	0.423
C2	*Lolium multiflorum*	0.019	0.135	<LOD	<LOD	<LOD	0.154
E1	*Lolium multiflorum*	<LOD	0.028	<LOD	<LOD	<LOD	0.028
E2	*Medicago sativa*	<LOD	<LOD	<LOD	<LOD	<LOD	–
E3	*Medicago sativa*	<LOD	1.600	<LOD	<LOD	<LOD	1.600
E4	Spontaneous grasses	<LOD	<LOD	<LOD	<LOD	<LOD	–
D1	*Lolium multiflorum*	<LOD	0.069	<LOD	<LOD	<LOD	0.069
D2	*Lolium multiflorum*	<LOD	1.090	<LOD	<LOD	<LOD	1.090
D3	*Medicago sativa*	<LOD	0.014	<LOD	<LOD	<LOD	0.014

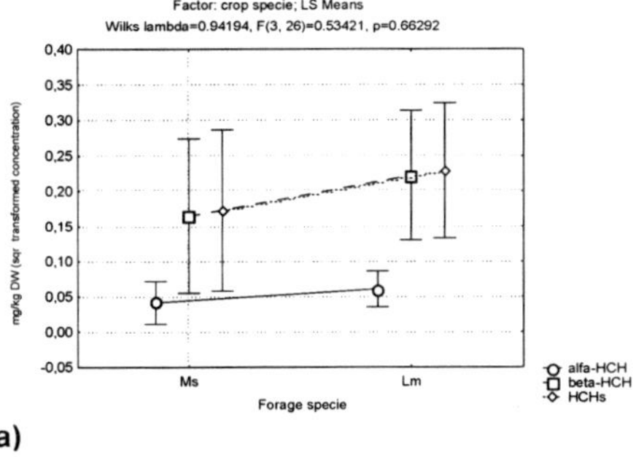

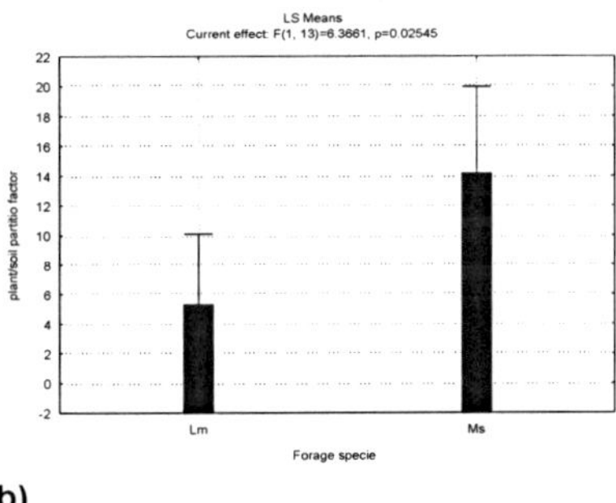

a)   b)

Figure 4. (a) LS Means of HCH concentration of soil samples (square rot transformed): effect of the species cropped where samples were collected. Vertical bars denote 0.95 confidence intervals. (b) LS means of HCH apparent plant/soil partition factor observed for different crop species. Lm = *Lolium multiflorum* (Graminae), Ms = *Medicago sativa* (Leguminosae)

3.2. MILK CONTAMINATION BY B-HCH

Data on milk contamination by β-HCH (the only isomer found), grouped for sampling date and farm, are reported in Table 3. Except the Farm E, where a detectable amount of β-HCH was never observed, in other farms only the β-HCH was found in 85.6% of the samples analyzed.

As far as the farm A and farm B is concerned, in only few cases β-HCH levels in milk exceeded the EU limit (0.0030 mg β-HCH/kg). Moreover, average data and variability over the entire survey, were comparable, whereas a generalised very high milk contamination was observed in farm D and E. Within each farm and sampling date, the variability of milk β-HCH content appeared high with a SDR up to near 200% (Farm D, 04/08/05). Similar variability has been reported by other authors about cow and buffalo milk contaminated by HCHs among farms (Smit, 1988; Battu et al., 1989) and animals (Van den Hoek et al., 1975; Smit, 1988; Sitarska et al., 1995).

TABLE 3. Milk contamination (means ± SD) by β-HCH (μg β-HCH/kg)

Sampling date	Farm A (n = 3)[a]	Farm B (n = 3)[a]	Farm C (n = 9)[a]	Farm D (n = 9)[a]	Farm E (n = 9)[a]
27/07/05– 04/08/05	0.0033 ± 0.0029	0.0019 ± 0.0032	nd	0.0055 ± 0.0108	0.0440 ± 0.0376
18/08/05	0.0013 ± 0.0004	0.0011 ± 0.0002	nd	0.0009 ± 0.0006	0.0708 ± 0.0922
09/09/05	0.0010 ± 0.0007	0.0008 ± 0.0014	nd	0.0023 ± 0.0035	0.1069 ± 0.0739
29/09/05	0.0018 ± 0.0003	0.0035 ± 0.0006	nd	0.0022 ± 0.0011	0.2845 ± 0.1867
Overall mean	0.0020 ± 0.0015	0.0018 ± 0.0019	–	0.0027 ± 0.0054	0.1266 ± 0.1431

[a]Number of subjects
nd = not detectable (<LOD)

TABLE 4. Nested ANOVA performed on data of milk contamination by β-HCH

	Factor	SS	df	MS	F	p
1	Sampling date[a]	5.745	3	1.915	4.946	0.0038
2	Dairy farm (nested in sampling date)[b]	13.146	9	1.461	3.772	0.0008
3	N. of lactations (nested in 2)[c]	1.543	1	1.543	3.985	0.0502
4	Lactation phase (nested in 3)[d]	5.267	4	1.317	3.401	0.0140

[a]Four levels. Date codes: 1 = 27/07/05–04/08/05, 2 = 18/08/05, 3 = 09/09/05, 4 = 29/09/05
[b]Four levels. Farms A, B, D and E. Farm C was not included in the analysis due to lacking of positive cases
[c]Two levels. 1 = lactation number ≤ 2, 2 = lactation number >2
[d]Three levels. 1 = 0–100 days of the cattle lactation curve, 2 = 100–200 days of the cattle lactation curve, 3 = 200–300 days of the cattle lactation curve

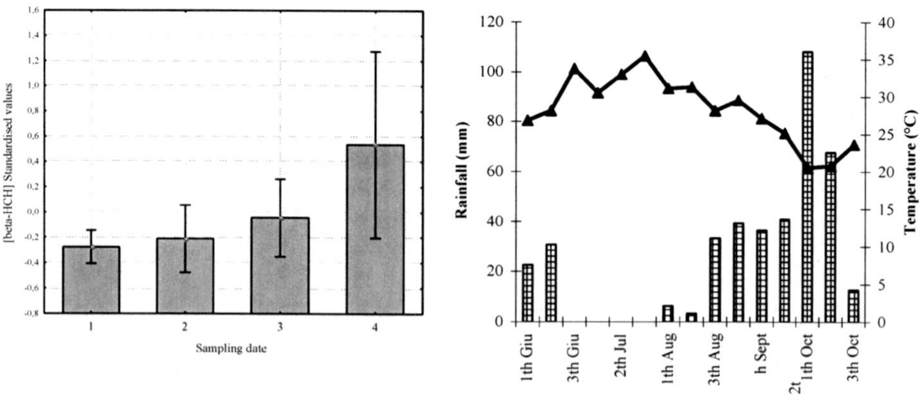

Figure 5. LS Means of β-HCH concentration in milk (standardized values). Current effect sampling date: F(3, 63) = 4.9461, p = 0.00380. Type III decomposition. Vertical bars denote 0.95 confidence intervals. Date codes: 1=27/07/05–04/08/05, 2=18/08/05, 3=09/09/05, 4=19/09/05   *Figure 6.* Means values per decade of air temperature (solid line) and rainfall (vertical bars) recorded at the A.R.S.I.A.L. Anagni Meteorological Station during the year 2005

To test the significance of differences a Nested Design ANOVA was performed on standardised data considering also parity and lactating phase as factors (Table 4). Date of sampling showed a strong effect (Figure 5) with a generalised increase of contamination during the course of the study ($p < 0.01$). Little information is available in literature about seasonal variations of OCPs residues in bovine milk. John et al. (2001), during a 4 year study in Jaipur City (Rajasthan, India), suggested as a general rule that seasonal climatic conditions may affect the level of milk contamination by OCPs including HCH, in bovine and buffalo milk.

Those authors found a general increase of OCPs residues in milk during the winter season when wind, heat and rain are usually very low in the Rajasthan region. They reported that these climatic conditions may affect environmental contamination, reducing the spreading of HCH after pesticide applications. Our findings suggest an opposite situation in agreement with Laquet et al. (1974) which, in a 4 years study on cow milk sold in France, found a higher concentration of HCH (isomers $\alpha + \beta + \gamma$) in during fall and spring, than during summer or winter season. In particular, their study shown a minimum for HCH contamination of milk during August 1971 and 1972 and an increase in HCH levels in September and October (roughly +40% in 1971 and +90% in 1972). A similar behaviour was observed for the pesticide heptachlor-epoxide. Air temperature and rainfall recorded during the control period in the contaminated area, are reported in Figure 6. From July to September 2005, rainfall per decade increased from 0.0 to 41.0 mm, while air temperature, as mean of 10 daily peak, decreased from 35.5°C to 25.4°C. Wide changes of local climatic conditions can affect the HCH isomers migration. Supporting this hypothesis, several studies showed that HCH isomers migration among different environmental compartments is mostly driven by the air temperature (Mackay, 1991; Komp and McLachlan, 1997; Kelly and Gobas, 2003), wind speed and direction (Glotfelty et al., 1984; Cleemann et al., 1995; Wittich and Siebers, 2002), air humidity and rainfall (Samuel and Pillai, 1990; Hippelein and MacLachlan, 2000). Lactation phase, showed a significant effect (p < 0.05) on milk

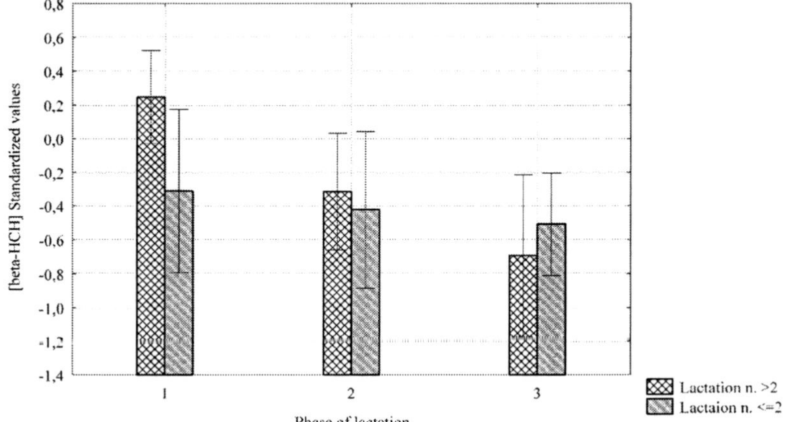

Figure 7. LS Means of β-HCH concentration in milk (standardized values). Current effect phase of lactation (Number of Lactations): F(4, 63) = 3.4006, p = 0.01395. Type III decomposition. Vertical bars denote 0.95 confidence intervals. Lactating stage codes: 1 = 0–100 days of the cattle lactation curve, 2 = 100–200 days of the cattle lactation curve, 3 = 200–300 days of the cattle lactation curve

contamination by β-HCH (Figure 7). Levels of β-HCH in milk were found to be higher during the first 100 days in milk than in the rest of lactation. In the firs part of lactation, due to the negative energy balance (De Vries et al., 1999; Banos et al., 2005.), cows must mobilize fat deposits and, consequently, all lipophilic substances stored in fats.

This was particularly evident in cows with three or more lactations due to their higher peak of lactation (Wood, 1967). Fries (1977) and more recently Dixon et al. (2000)[5] pointed out that the concentrating effect of lipophilic contaminants in milk due to weight loss and fat mobilisation is stronger than the effect of concomitant diluting effect due to the higher milk production.

3.3. B-HCH IN BLOOD SERUM

Blood samples collected from eight cows from farm E during three samplings in August 2005 and September 2005, resulted highly positive to β-HCH (Table 5). As reported above, see Table 1, this farm showed the highest degree of milk contamination by β-HCH. Surprisingly, the average level reached by this contaminant in blood, decreased from the beginning to the end of the experimental period. In contrast, β-HCH level in milk increased. This may be the consequence of a severe deviation from equilibrium in β-HCH partitioning among fat depots in animal adipose tissue, serum and milk (Waliszewski et al., 2004). However considering medians, such differences of β-HCH milk content were not observed (Table 5) suggesting that the average trend may be due to a distribution far away from normality. A quite high degree of variability in blood serum β-HCH level was observed. Similar heterogeneity has been reported in other researches mainly focused on peoples chronically exposed to HCHs (Otero et al., 1997; Waliszewski et al., 2000, Karmaus et al., 2005; Waliszewski et al., 2004; Siddiqui et al., 2005; Thomas et al., 2005).

The relationship between blood serum and milk β-HCH levels is reported in Figure 8. Using all data, there was no relationship (r^2 = 0.0124, p = 0.446) between blood and milk (Figure 8a). Excluding two cases falling outside the 95% confidence limit of regression, the relationship becomes significant (r^2 = 0.8519, p = 0.009) (Figure 8b). Moreover, such an exclusion of outlier cases, leaded to a not significance of the intercept value (p = 0.065), adding robustness to the linear relation. As reported above, cows from farm C were always negative for β-HCH in milk throughout the study. As a presumable negative control, three blood samples were analyzed to determine β-HCH. In two cases a trace (below the LOQ) of β-HCH were found (estimated level 0.0003 µg β-HCH/ml). As pointed out by others (Gupta et al., 1978; To-Figueras et al., 1997; Otero et al., 1997; Waliszewski et al., 2004) our results suggest a high sensibility of blood as an early indicator of OCP exposure.

---

[5] Dixon, F., Diment, A., Ambroce, K., 2000, Reducing pesticide residues in cattle. Farmnote 19/2000. Department of Agriculture, Western Australia, http://www.agric.wa.gov.au/content/PW/CHEM/FN2005_OCRESIDUES_CATTLE.PDF

TABLE 5. Descriptive statistics of β-HCH contamination of blood serum samples collected in farm E

Sampling date	N	Positive cases (%)	Mean[a]	SD	SDR (%)	Range	Median	25th[a]	75th[a]
Farm E									
04/08/05	8	88	0.0329	0.0443	134	0.116	0.0110	0.001	0.061
09/09/05	8	100	0.0141	0.0151	107	0.049	0.0105	0.005	0.016
29/09/05	8	100	0.0115	0.0049	43	0.016	0.0120	0.009	0.014
All data	24		0.0194	0.0282	142	0.116	0.0110	0.002	0.017
Farm C									
04/08/05	1	100	0.0003[b]						
09/09/05	1	100	0.0003[b]						
29/09/05	1	0	<LOD						

[a]Data expressed as µgβ-HCH/ml
[b]Data below LOQ, estimated values

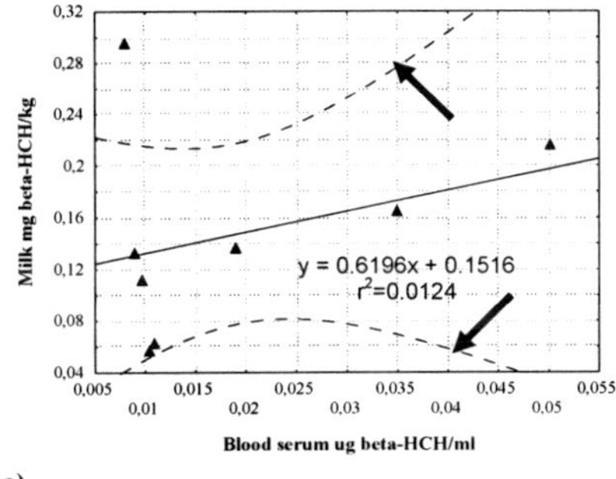

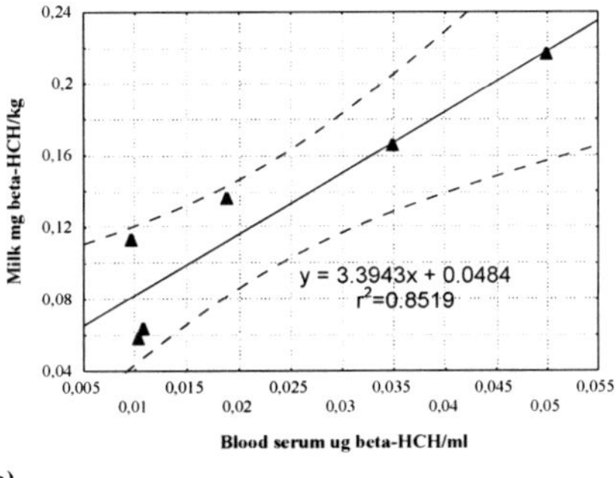

Figure 8. Farm E, mean β-HCH level found in cow blood serum *vs.* milk content: (a) all cases, two cases indicated by an arrow were outside of the 95% confidence limits of regression, (b) regression without the outlier cases

4. Conclusion

The case study on HCHs pollution in a rural area of central Italy, highlights a high degree of complexity, due to interactions of chemical compounds with components of agro-ecosystem, such as soil, climate, vegetation and farming practices. Contamination of farm lands by POPs poses serious problems for what concerns food safety, health of human population living in rural areas and also for animal health. The degree of contamination by POPs all around the world requires a common effort for a better comprehension of risk factors and for a correct planning of preventive measures. Much work remains to do in the fields of risk management and risk communication during episodes of crisis. Further studies are needed for a better understanding of agro-environmental fate and impact of POPs, of toxic effects of acute and chronic exposure in humans and animals, and of feasibility and effectiveness in practical conditions of bio-remediation systems.

Acknowledgements

Research granted by Province of Rome.

References

ATSDR, 2003, Toxicological profile for alpha-, beta-, gamma-, and delta-hexachlorocycloexane, Agency for Toxic Substances and Disease Registry. US Department of Health and Human Services, Georgia, USA, 306 pp.

Ayas, Z., Barlas, N., Kolankaya, D., 1997, Determination of organochlorine pesticide residues in various environments and organisms in Goksu Delta, Turkey, *Acquat. Ecol.*, 39: 171–181.

Barrie, L.A., Gregor, D., Hargrave, B., Lake, R., Muir, D.C.G., Shearer, R., 1992, Arctic contaminants: sources, occurrence and pathways, *Sci. Total Environ.*, 122: 1–71.

Banos, G., Coffey, M.P., Brotherstone, S., 2005, Modeling daily energy balance of dairy cows in the First Three Lactations, *J. Dairy Sci.*, 88: 2226–2237.

Battu, R.S., Singh, P.P., Joia, B.S., Kalra, R.L., 1989, Contamination of bovine (Buffalo, Bubalus bubalis (L.)) milk from indoor use of DDT and HCH in malaria control programmes, *Sci. Total Environ.*, 86(3): 281–287.

Breivika, K.U., Pacynaa, J.M., Munch, J., 1999, Use of α-, β- and γ-hexachlorocyclohexane in Europe, 1970–1996, *Sci. Tot. Environ.*, 239: 151–163.

Carter, R.H., 1947, Estimation of DDT in Milk by Determination of Organic Chlorine, *Ind. Eng. Chem. Anal.*, 19: 54–62.

Carter, R.H., Hubanks, P.B., Poos, F.W., Moore, L.A., Ely, R.B., 1953, The toxaphene and chlordane content of milk from cows receiving these materials in their feed, *J. Dairy Sci.*, 36: 1172–1177.

Clark, T., Clark, K., Paterson, S., Nostrom, R., Mackay, D., 1988, Wildlife monitoring, modelling and fugacity. *Environ. Sci. Technol.*, 22: 120–127.

Cleemann, M., Poulsen, M.E., Hilbert, G., 1995, Deposition of Lindane in Denmark, *Chemosphere*, 11: 2039–2049.

Concha-Grana, E., Turnes-Carou, M.I., Muniategui-Lorenzo, S., Lopez-Mahia, P., Prada-Rodriguez, D., Fernandez-Fernandez, E., 2006, Evaluation of HCH isomers and metabolites in soils, leachates, river water and sediments of a highly contaminated area, *Chemosphere*, 64: 588–595.

Covaci, A., Hura, C., Schepens, P., 2001, Selected persistent organochlorine pollutants in Romania, *Sci. Total Environ.*, 280: 143–152.

De Vries, M.J., Van Der Beek, S., Kaal-Lansbergen, L.M.T.E., Ouweltjes, W., Wilmink, J.B.M., 1999, Modeling of energy balance in early lactation and the effect of energy deficits in early lactation on first detected estrus postpartum in dairy cows, *J. Dairy Sci.*, 82: 1927–1934.

DIN – Deutsches Institut fur Normung e.V.- Norma 32645 (1994), *Chemical analysis; decision limit; detection limit and determination limit; estimation in case of repeatability; terms, methods, evaluation.* Beuth-Verlag, Berlin, Germany, 24 pp.

FAO, 2005, Seventh FAO consultation on prevention and disposal of obsolete, banned and unwanted pesticide stocks, Food and Agriculture Organization of the United Nations, Rome Italy, 152 pp.

Frazar, C., 2000, *The bioremediation and phytoremediation of pesticide-contaminated sites.* US Environmental Protection Agency, Office of Solid Waste and Emergency Response Technology Innovation Office, Washington, DC, 180 pp.

Fries, G.F., 1977, The kinetics of halogenated hydrocarbons retention and elimination in dairy cattle. In: Fate of pesticides in large animals, G.W. Ivie and H.W. Dorough (Eds.). Academic Press, Inc., 159–173.

Galassi, S., Viganò, L., Sanna, M., 1996, Bioconcentration of organochlorine pesticides in rainbow trout caged in the river Po, *Chemosphere*, 32(9): 1729–1739.

Garmouna, M., Poissant, L., 2004, Occurrence, temperature and seasonal trends in α- and γ-HCH ina air (Québec, Canada), *Atmos. Environ.*, 38: 369–382.

Goltfelty, D.E., Taylor, A.W., Turner, B.C., Zoller, W.H., 1984, Volatilization of surface-applied Pesticides from fallow soil, *J. Agric. Food Chem.*, 32: 638–643.

Gupta, R.C., Karnik, A.B., Nigan, S.K., Kashyap, S.K., 1978, Comparative laboratory evaluation of some reported methods for the determination of DDT and BHC insecticides in human blood, *Analyst*, 103: 723–727.

Karmaus, W., Brooks, K.R., Nebe, T., Written, J., Obi-Osius N., Kruse, H., 2005, Immune function biomarkers in children exposed to lead and organochlorine compounds: a cross-sectional study, *Environmental Healt*, 14(4): 5.

Kelly, B.C., Gobas, F.A.P.C., 2003, An arctic terrestrial food-chain bioaccumulation model for persistent organic pollutants, *Environ. Sci. Technol.*, 37(13): 2966–2974.

Komp, P., McLachlan, M.S., 1997, Interspecies variability of the plant/air partitioning of polychlorinated biphenyls, *Environ. Sci. Technol.*, 31: 2944–2948.

Hans, R.K., Farooq, M., Suresh Babu, G., Srivastava, S.P., Joshi, P.C., Viswanathan, P.N., 1999, Agricultural produce in the dry bed of the River Ganga in Kanpur, India – New source of pesticide contamination in human diets, *Food Chem. Toxicol.*, 37: 847–852.

Harding, G.C., LeBlanc, R.J., Vass, W.P., Addison, R.F., Hargrave, B.T., Pearre, S., Dupuis, A., Brodie, P.F., 1997, Bioaccumulation of polychlorinated biphenyls (PCBs) in the marine pelagic food web, based on a seasonal study in the southern Gulf of St. Lawrence, 1976–1977, *Mar. Chem.*, 56(3–4): 145–179.

Harrad, S., Mao, H., 2004, Atmospheric PCBs and organochlorine pesticides in Birmingham, UK: concentrations, sources, temporal and seasonal trends, *Atmos. Environ.*, 38: 1437–1445.

Hayes, W.J., 1982, Pesticides studied in man. Williams & Wilkins, Baltimore, MD, 372 pp.

Hippelein, M., McLachlan, S., 2000, Soil/air partitioning of semivolatile organic compounds. 2. Influence of temperature and relative humidity, *Environ. Sci. Technol.*, 34: 3521–3526.

Iwata, H., Tanabe, S., Sakal, N., Tatsukawa, R., 1993, Distribution of persistent organochlorines in the oceanic air and surface seawater and the role of ocean on their global transport and fate, *Environ. Sci. Technol.*, 27: 1080–1098.

John, P.J., Bakore, N., Bhatnagar, P., 2001, Assessment of organochlorine pesticide residue levels in dairy milk and buffalo milk from Jaipur City, Rajasthan, India, *Environ. Int.*, 26: 231–236.

Jolly, D.W., 1967, Organochlorine Pesticides and Farm Livestock, *Proc. R. Soc. Med.*, 60: 25–28.

Langenhoff, A.A.M., Staps, J.J.M., Pijls, C., Alphenaar, A., Zwiep, G., Rijnaarts, H.H.M., 2002, Intrinsec and stimulated *in situ* biodegradation of hexachlorocycloexane (HCH), *Water Air Soil Poll.*, 2: 171–182.

Laquet, F., Goursaud, J.G., Casalis, J., 1974, Les residues de pesticide organochlores dans les laits animaux et humains, *Ann. Fals. Exp. Chim.*, 716(67): 217–239.

Losada, A., Fernàndez, N., Diez, M.J., Teràn, M.T., Garcìa, J.J., Sierra, M., 1996, Organochlorine pesticide residues in bovine milk from Léon (Spain), *Sci. Total. Environ.*, 181: 133–135.

Mackay, D., 1991, *Illustrated handbook of physical-chemical properties and environmental fate for organic chemicals: pesticide chemicals.* CRC Press-Lewis, 812 pp.

Mallatou, H., Pappas, C.P., Albanis, T.A., 2002, Behaviour of pesticides Lindane and methyl parathion during manufacture, ripening and storage of feta cheese, *Int. J. Dairy Technol.*, 55(4): 211–216.

Manz, M., Wenzel, K.-D., Dietze, U., Schuurmann, G., 2001, Persistent organic pollutants in agricultural soils of central Germany, *Sci. Total Environ.*, 277: 187–198.

McGrath, D., 1995, Organic micropollutant and trace element pollution of Irish soils, *Sci. Total Environ.*, 164: 125–133.

Meijer, S.N., Halsall, C.J., Harner, T., Peters, A.J., Ockenden, W.A., Johnston, A.E., Jones, K.C., 2001, Organochlorine pesticide residues in archived UK soil, *Environ. Sci. Technol.*, 35: 1989–1995.

Metcalf, R.L., 1955, Organic insecticides, their chemistry and mode of action. Interscience Publishers, New York, 392 pp.

Misra, K., Satinder Brar, K., Verma, M., Tyagi, R.D., Trivedy, R.K., Sharma, S., 2007, Lindane – A contaminant of global concern, *J. Ind. Poll. Contr.*, 23(1): 169–187.

Naso, B., Zaccaroni, A., Perrone, D., Ferrante, M.C., Severino, L., Stracciari, G.L., Lucidano, A., 2004, Organochlorine pesticides and polychlorinated biphenyls in European roe deer *Capreolus capreolus* resident in a protected area in Northern Italy, *Sci. Total Environ.*, 328: 83–93.

Nerin, C., Martinez, M., Cacho, J., 1991, Distribution of hexachlorocyclohexane isomers from several sources, *Toxicol. Environ. Chem.*, 35: 125–135.

Nikanorov, A.M., Korotova, L.G., Klimenko, O.A., 2007, Estimation of long-term trends in the discharge of organochlorine pesticides into seas by rivers of Russia, *Water Resour.*, 34(4): 415–422.

Otero, R., Santiago-Silva, M., Grimalt, J.O., 1997, Hexachlorocyclohexane in human blood serum, *J. Chrom. A*, 778: 87–94.

Samuel, T., Pillai, M.K.K., 1991, Impact of repeated application on the binding and persistence of [14C]-DDT and [14C]-HCH in a tropical soil, *Environ. Poll.*, 74(3): 205–216.

Shapiro, S.S., Francia, R.S., 1975, An approximate analysis of variance test for normality, *J. Am. Stat. Assoc.*, 67: 215–216.

Sharma, H.R., Kaushik, A., Kaushik, C.P., 2007, Pesticide residues in bovine milk from a predominantly agricultural state of Haryana, *Ind. Environ. Monit. Assess.*, 129: 349–357.

Shen, L., Wania, F., Lei, Y.-D., Teixeira, C., Muir, D.C.G., Bidleman, T.F., 2005, Atmospheric distribution and long-range transport behaviour of organochlorine pesticides in North Shepherd, Report, USA.

Sheperd, J.B., Moore, L.A., Carter, R.H., Poos, F.W., 1949, The effect of feeding alfalfa hay containing DDT residue on the DDT content of cow's milk, *J. Dairy Sci.*, 32: 549–555.

Skrbic', B., Durisic-Mladenovic, N., 2007, Principal component analysis for soil contamination with organochlorine compounds, *Chemosphere*, 68: 2144–2152.

Siddiqui, M.K.J., Anand, M., Mehrotra, P.K., Sfrangi, R., Mathur, N., 2005, Biomonitoring of organochlorines in women with benign and malignant breast disease, *Environ. Res.*, 98: 250–257.

Sitarska, E., Kluciniski, W., Faundez, R., Duszewska, A.M., Winnicka, A., Géralczyk, K., 1995, Concentration of PCBs, HCB, DDT, and HCH isomers in the ovaries, mammary gland, and liver of cows, *Bull. Environ. Contam. Toxicol.*, 55: 865–869.

Smit, L.E., 1988, Survey of pesticides in milk, *South Afr. J. Dairy Sci.*, 20(1): 37–38.

Simonich, S.L., Hites, R.A., 1995, Global distribution of persistent organochlorine compounds, *Science*, 269(5232): 1851–1854.

Spiric, A., Saicic, S., 1998, Monitoring chlorinated pesticides and toxic elements in tissues of food-producing animals in Yugoslavia, *J. AOAC Int.*, 81(6): 1240–1243.

Tanabe, S., Tatsukawa, R., Kawano, M., Hidaka, H., 1982, Global distribution and atmospheric transport of chlorinated hydrocarbons: HCH (BCH) isomers and DDT compounds in the Western Pacific, Eastern Indian and Antarctic Ocean, *J. Oceanogr. Soc. Japan*, 38: 137–148.

Tanabe, S., Hidaha, H., Tatsukawa, R., 1983, PCB and Chlorinated Hydrocarbon pesticides in the Antartic atmosphere and hydrosphere, *Chemosphere*, 12: 277–288.

Tanabe, S., Iwata, H., Tatsukawa, R., 1994, Global contamination by persistent organochlorines and their ecotoxicological impact on marine mammals, *Sci. Total Environ.*, 154: 163–177.

Thomas, G.O., Wilkinson, M., Hodson., S., Jones, K.C., 2005, Organohalogen chemicals in human blood from the United Kingdom, *Environ. Poll.*, 141(1): 30–41.

To-Figueras, J., Sala, M., Otero, R., Barrot, C., Santiago-Silva, M., Rodamilans, M., Herrero, C., Grimalt, J., Sunyer, J., 1997, Metabolism of hexachlorobenzene in humans: association between serum levels and urinary metabolites in a highly exposed population, *Environ. Health Perspect.*, 105(1): 78–83.

UNEP, 1995, *Towards global action. International experts meeting on persistent organic pollutants*, Meeting Background Report. Vancouver, Canada, June 4–8, Meeting Background Report, 201 pp.

US-EPA, 1994, *Deposition of air pollutants to the Great Waters. First Report to congress*, United States Environmental Protection Agency, EPA-453rR-93-055, 1994, 164 pp.

Van Barneveld, R.J., 1999, Physical and chemical contaminants in grains used in livestock feeds, *Aust. J. Agric. Res.*, 50: 807–823.

Van Den Hoek, J., Salverda, M.H., Tuinstra Th., L.G.M., 1975, The excretion of six organochlorine pesticides into the milk of the dairy cow after oral administration, *Neth. Milk Dairy. J.*, 29(1): 66–78.

Verna, A., Pillai, M.K.K., 1991, Bioavailability of soil-bound residues of DDT and HCH to certain plants, *Soil Biol. Biochem.*, 23(4): 347–351.

Waliszewski, S.M., Pardio, V.T., Waliszewski, K.N., Chantiri, J.N., Aguirre, A.A., Infanzòn, R.M., Riveire, J., 1997, Organochlorine pesticide residues in cow's milk and butter in Mexico, *Sci. Total. Environ.*, 208: 127–132.

Waliszewski, S.M., Aguirre, A.A., Infanzón, R.M., Siliceo, J., 2000, Carry-over of persistent organochlorine pesticides through placenta to fetus, *Salud Publica Mex.*, 42: 384–390.

Waliszewski, S.M., Carvajal, O., Infranzon, R.M., Trujillo, P., Hart, M.M., 2004, Copartition ratios of persistant organochlorine pesticides between adipose tissues and blood serum lipid, *Bull. Environ. Contam. Toxicol.*, 73: 732–738.

Wania, F., Mackay, D., 1996, Tracking the distribution of persistent organic pollutants. Control strategies for these contaminants will require a better understanding of how they move around the globe, *Environ Sci. Technol.*, 30(9): 390A–396A.

Wenzel, K.-D., Hubert, A., Weissflog, L., Kuhne, R., Popp, P., Kindler, A., Schuurmann, G., 2006, Influence of different emission sources on atmospheric organochlorine patterns in Germany, *Atmos. Environ.*, 40: 943–957.

Wittich, K.-P., Siebers, J., 2002, Aerial short-range dispersion of volatilized pesticides from an area source, *Int. J. Biometeorol.*, 46: 126–135.

Wood, P.D.P., 1967, Algebric model of the lactation curve in cattle, *Nature*, 216: 164–165.

Word Bank CIDA, 2001, Persistent Organic Pollutants and the Stockholm Convention: a resource guide. Washington DC, USA, 185 pp.

Yeo, H.-G., Choi, M., Sunwooa, Y., 2004, Seasonal variations in atmospheric concentrations of organochlorine pesticides in urban and rural areas of Korea, *Atmos. Environ.*, 38: 4779–4788.

ASSESSING THE EXTENT OF POLLUTANT ACCUMULATION IN THE ANIMAL FOODS AND BLOOD OF INDIVIDUALS INHABITING THE AZGYR TEST BASE AREA

USEN KENESARIYEV, NIYAZALY ZHAKASHOV, IVAN SNYTIN, MEIRAM AMRIN, YERZHAN SULTANALIYEV
S.D.Asfendiyarov Kazakh National Medical University
Chair of Common Hygiene and Ecology, Almaty

Abstract: The studies have been undertaken in the Azgyr nuclear test base located in the western part of Atyrau Region, Republic of Kazakhstan, where 17 subsurface explosions had been set off from 1966 through 1979. We have studied the natural environments (soil, plants, food), biological environment of the wildlife and the human beings. Radionuclides intake in foodstuff in the Azgyr test base area was due to their aerogenic precipitation on the pasture grass that is especially significant in the period of active grazing. Multiple subsurface nuclear explosions in the salt domes of the test base have resulted in the increased water hardness and increased contents of lead and cadmium which exceed the limits. Concentrations of cobalt in the soils of inhabited localities were 21 times higher than normal, cadmium – 6 times higher and lead – 10 times. The content of cobalt in the organs of small cattle (liver, kidneys, heart, and lungs) was 2.5 times higher than MCL, and five times higher than normal – in milk. High level of lead, cadmium, and cobalt and deficiency of important microelements (zinc, copper, and iron) was observed in the blood samples of individuals inhabiting the region.

Keywords: Radionuclides, heavy metals, biological environments

1. Introduction

The purpose of the study was to assess pollutant accumulation in the animal foods and blood of individuals inhabiting the Azgyr nuclear test base area.

The Azgyr nuclear test base is located in the arid zone on the Bolshoy Azgyr salt dome in the vicinity of Azgyr settlement, Kurmangazy District, Atyrau Region.

Seventeen subsurface nuclear explosions had been set off from 1966 through 1979 in 10 wells at the depths of 165–1,500 m to create underground cavities.

2. Materials and Methods

We have examined and tested natural environments (soil, plants, foods), biological environments of the wildlife and the human blood samples in the Azgyr test base area. We have used hygienic, statistical and epidemiological methods.

3. Results and Discussion

3.1. SOIL AND GRASS CONTAMINATION

Now, 41–28 years passed since the first and the last explosions at the Azgyr test base area. According to the National Nuclear Center and the Committee for Hydrometeorology of Kazakhstan (1995), the soil is still rather contaminated with Cs-137 at the depth of 5 cm, the contamination ranging between 80 and 13,600 mCu/km^2, with the global capacity of 65 mCu/km^2.

The soils in the Azgyr test base area are mainly sandy therefore the migration from the soil into plants took place (see Table 1).

TABLE 1. Total beta-activity and gamma-background of natural environments in the Azgyr test base area

Radiological Analysis	Balkuducks		Suyunducks	
Soil	3 km from A-2	5 km from A-2	3 km from A-2	5 km from A-2
γ-Background (mcr/h)	13–14	13–14	14	15
Total β-activity (Cu/kg)	3.72×10^{-3} 5.09×10^{-8}	3613×10^{-3}	14.11×10^{-8}	3.2×10^{-8}
Radiation standards-99	$5.5 \ \times 10^{-8}$			
Plants (grass, rush, hay, motley grasses)	3 km from A-2	5 km from A-2	3 km from A-2	5 km from A-2
γ-Background (mcr/h)	13–14	13	13	14–15
Total β-activity (Cu/kg)	0.21×10^{-8} 0.99×10^{-8}	0.51×10^{-8}	1.03×10^{-8}	1.95×10^{-8}
Radiation standards-99	2×10^{-8}			

Gamma-background and total β-activity of the soil didn't exceed RS-99 except for exceeding RS-99 of soil β-activity by 2.5 times within a radius of 3 km from A-2 site. Gamma-background and total β-activity of the plants were in line with RS-99 (Table 1).

The transport factor of Cs-137 (proportion of its concentrations in the solid grass and soil) was in the range of 0.2 ± 0.003 (in autumn) to 1.5 ± 0.2 (in spring). The similar data were observed for Semipalatinsk test base.

3.2. WATER CONTAMINATION

The presence of radionuclides in the water bodies of the Azgyr area was due to direct sedimentation both on reservoir surface and mainly to the washout of radioactive substances from the soil with melted snow and rain waters.

TABLE 2. Radioactive composition of water in water sources in the Azgyr test base area (in km/h)

Location of sampling	Distance between sites and settlements	Radionuclide content	
		Sr-90	Cs-137
Surfaces of water sources (site A-9)	20–25 km to the north of Assan and Azgyr settlements	Below the method sensitivity threshold (BST)	BST
Shallow water source (site A-4)	20–25 km to the north of Assan and Azgyr settlements	BST	BST
Inspection well (site A-5)	20–25 km to the north of Assan and Azgyr settlements	4.2×10^{-12}	BST
Balkuduck settlement (drinking water from the well)	10.5 km to the south of A-2 site	BST	BST
Permissible activity level (PAL), RS-99	–	1.0×10^{-10}	5.0×10^{-10}

It was worth mentioning that the Azgyr test base was exposed to radiation pollution for many years after the last explosion since the operations for the nuclear site were laid up and cultivated. Also, radioactive waste burial was started much later – 10 years after the explosion. Thus, radioactive waste contributed to the migration of radionuclides in surface water and groundwater (Table 2).

The presence of Sr-90 and Cs-137 in the water of the inspection well of A-5 site located in 30 km from Assan and Azgyr settlements was the evidence of the water stratum being contaminated with the products of nuclear explosion (Table 2). Thus, the content of Sr-90 in the groundwater amounted to 4.2×10^{-12} (with the norm being 1.0×10^{-10} km/h). In this connection, one should not exclude the possibility that groundwater at other sites was exposed to radioactive pollution as well. But since no inspection wells were available, these studies had not been carried out.

Of special importance was the radionuclide composition of drinking water in the Azgyr area as compared with the Kapustin Yar test base (Table 3).

TABLE 3. Comparative content of radionuclides in drinking water in the areas of the Azgyr and Kapustin Yar test bases for 2000–2003

Radionuclides Bq/kg	PAL	2000 Azgyr	2001 Azgyr	2002		2003 Azgyr
				Azgyr	Kapustin Yar	
Ra-226	0.5	0.126–0.145	0.068–0.074		4.7 ± 0.4 8.8 ± 1.3	0.074–0.106
Sr-90	5.0	BST	BST	BST	BST	0.04
Cs-137	11.0	BST	BST	BST	28 ± 0.1 3.7 ± 0.2	BST
U (mcg/l)		0.069	0.074–0.106	0.68		0.922

BST – below sensitivity threshold

Radionuclides in drinking water did not exceed the permissible activity levels. A low content of Ra-226 and Cs-137 radionuclides in the Azgyr test base area could be observed as compared to the Kapustin Yar. This was accounted for by the fact that in the period from 1957 to 1962 ten high-altitude nuclear explosions had been set off with the yield of 1.2–300 kg, the anomaly length being about 180 km and the width – about 50 km, with the content of Cs-137 zone exceeding the average global non-point activity by 1.5–2.8 times.

3.3. ANIMAL PRODUCT CONTAMINATION

According to the Atyrau Regional Sanitary and Epidemiological Station, anthropogenous radionuclides (Sr-90 and Cs-137) were also found in milk and meat (Table 4).

Radionuclides in foods did not exceed permissible levels and were in lower quantity than in Austrian food after Tchernobyl fallout (Schwaiger et al., 2004). The downward trend of radionuclide content was observed in 2003 as compared with the years of 2000–2001: in meat Ra 226 – from 0.175 to 0.103 Bq/kg, Sr 90 – from 0.222 to 0.103, Cs 137 – from 0.231 to 0.133, Pb 210 – from 0.149 to 0.048, in milk Ra 226 – from 0.229 to 0.092, Sr 90 – from 0.166, Cs 137 – from 0.247 to 0.125, Pb 210 – from 0.165 to 0.055.

TABLE 4. Radionuclide composition (Bq/kg) of meat and milk in the Azgyr test base area for years 2000–2003

Years	Radionuclides	Meat	Milk
2000	Ra 226	0.135–0.175	0.096–0.229
	Sr 90	0.174–0.222	0.115–0.116
	Cs 137	0.208–0.231	0.118–0.205
	Pb 210	0.138–0.149	0.093–0.122
2001	Ra 226	0.087	0.059–0.096
	Sr 90	0.052	0.022–0.033
	Cs 137	0.121	0.048–0.099
	Pb 210	0.023	0.018–0.058
2002	Ra 226	0.103	0.077–0.092
	Sr 90	0.103	0.063–0.092
	Cs 137	0.133	0.096–0.125
	Pb 210	0.048	0.051–0.055
2003	Ra 226	0.103	0.077–0.092
	Sr 90	0.147	0.072–0.126
	Cs 137	0.114	0.151–0.244
	Pb 210	0.137	0.137–0.165
RS-99	Sr 90	50.0	25.0
	Cs 137	100.0	40.0

Total β-activity of meat of small cattle (SC) ranged between 1.67×10^{-9} and 2.279×10^{-9}, bones of SC- between 0.283×10^{-9} and 3.824×10^{-9}, which was less than the permissible level (1.6×10^{-8} Cu/kg).

3.4. CONSEQUENCES ON INHABITANTS

Multiple subsurface nuclear explosions in the salt domes of the Azgyr test base resulted in the increased hardness of drinking water in the inhabited localities that almost double the normal value (mineshafts), increased presence of lead – up to 3.0–6.3 MCL, and cadmium – up to 9.8 MCL. The chloride contents in the shallow waters in the vicinity of A-7 and A-9 sites ranged between 7 and 153 times of the normal value, lead content – from 1.0 to 3.0 MCL, the bottom slits of water bodies contained: lead – up to 240 MCL, cobalt – up to 12 MCL, and manganese – up to 37 MCL. The concentration of maximal cobalt in the soils of the inhabited localities reached 21 MCL, cadmium –6 MCL, and lead –10 MCL: in the organs of SC (liver, kidneys, heart, and lungs) maximal cobalt content reached 2.5 MCL, in milk – up to 5.0 MCL, and lead content – from 0.5 to 0.9 MCL.

In this connection whole blood tests were made among the inhabitants of the Azgyr test base for the content of heavy metals. As known, lead absorption depended on the age: 16% – at 27 years old and 1.3% – at 59 years old.

The increased content of lead in the blood of community individuals was observed among those aged between 15–40 years old: from 0.23 to 0.33 mg/l that is 1.6 time higher the permissible natural content in the adults (0.20 mg/l). This is due to the high content of lead in the soil, drinking water (3.0–10.0 MCL) and in the local produce (meat, cow milk). For example in polluted town from Italy (Naples), the lead concentration in blood varied between 0.088 and 0.137 mg/l (Amodio-Cocchieri, 1996). It should be mentioned that as the people grow older (40–49 and older), the lead content in the blood tends to decrease from 0.33 to 0.21 mg/l.

The highest content of lead in the blood (0.27–0.28 mg/l) was observed in the men aged 50–59 which is 1.4 times higher than the permissible physiological level among the adults (0.20 mg/l), that fact being confirmed by the scientific sources.

The largest buildup of cadmium in a body was observed in the age group of 30–49 years old and it continued until 50 and more. In the Azgyr test base area, the cadmium content in the blood of individuals inhabiting the settlements of Balkuduck, Azgyr and Konyrtereck was observed mainly in those aged 30–39 and 50–59 – from 0.20 to 0.025 mg/l, that was four times higher than the permissible levels and was confirmed by the scientific sources.

In the Azgyr test base area, the cadmium content in the blood of individuals – from 0.019 to 0.02 mg/l – was observed in the male individuals inhabiting the settlements of Balkuduck, Konyrtereck and Batyrbeck and aged 40 and older and was 3.8 times higher than the permissible level for humans (0.0052–0.006 mg/l). In comparison, the mean blood cadmium level for smokers in Sweden was 0.00112 mg Cd/l compared with 0.00087 mg Cd/l for non-smokers (Hallen et al., 1995). The high content of cadmium in the blood of individuals dwelling in the Azgyr area was associated with its increased content in the soil (up to 6 MCL) and foods (meat, milk) – up to 0.5–0.9 MCL.

High contents of cobalt in the Azgyr text base area were observed in the blood samples of children (aged under 15) – from 0.19 to 0.21 mg/l that was higher than the natural permissible levels for humans, this fact being confirmed by other authors.

The content of cobalt in the blood of the majority of adults inhabiting the Azgyr text base area was lower than in children and varied from 0.14 to 0.18 mg/l and the cobalt concentration didn't exceed the permissible maximum levels.

Strong deficiency of zinc – almost 2.8 times less than the normal value (6.0–8.0 mg/l) was observed in the blood samples of individuals inhabiting the Azgyr text base area. Its content varied from 1.38–1.77 (settlements of Batyrbeck and Konyrtereck) to 2.22–2.45 mg/l (settlements of Azgyr, Balkuduck, and Suyunduck). At the same time, the highest zinc content – 2.4–4.82 mg/l – was found in the blood samples in the reference settlement of Ganyushkino. The strongest zinc deficiency was found in the blood samples of post pubertal children (under 15 years old) – from 0.43 to 0.84 (settlements of Batyrbeck and Konyrtereck). The content of zinc in blood increased among juvenile individuals (15–18 years old) from 1.03 to 1.58 mg/l (settlements of Azgyr, Suyunduck, Batyrbeck and Konyrtereck).

In the Azgyr text base area the individuals aged 50–59, 60 and older show increased content of zinc – from 2.7 to 4.82 mg/l, with the lowest value in the female individuals – 1.72 mg/l on average, and in the male individuals – 2.56 mg/l. The normal level of zinc in serum was considered between 84 and 163 µg/100 ml in western countries (Fredricks et al., 1960).

Many ailments, both among children and adults, were characterized by the explicit exogenous copper deficiency in the body. The strongest copper deficiency was observed in the blood samples of children under 15 years old (0.13–0.37 mg/l) versus the adult individuals (0.23–0.51 mg/l). Normal copper in human blood was reported as 175.2 ± 14.3 µg/100 ml (Tefas et al., 1998).

In the Azgyr text base area iron deficiency was mainly observed in the juveniles (19–20 years old) and iron concentrations in the blood were the lowest – from 144.3 to 240.5 mg/l. (Azgyr and Batyrbeck settlements). The average content of iron in the blood of female individuals was lower (437.9–459 mg/l) than among the male individuals (485.8–605.2 mg/l). The normal content of iron in the whole blood was from 40 to 500 mg/l. As they age, male individuals tend to have the increasing content of iron in their blood and red blood cells. Thus, in the Azgyr text base area the content of iron in the blood in the individual age groups of 40–49 and 60 years and older increased to 605.66–856.18 mg/l. Male individuals aged 50–59, 60 and older slightly increased content of iron in their blood as compared with female individuals – on average, 587.4 mg/l versus 514.0 mg/l.

Explosions resulted in the increased radiation and chemical burden on the environment and the human body hence the increased daily salt load with the drinking water was observed for magnesium, chlorides and fluorides, total real daily load of lead upon a human's body was three times higher than the normal, and in case of radionuclides – up to 17.4 times.

Following the nuclear tests, deterioration in the basic morbidity rates of the local inhabitants was observed.

References

Amodio-Cocchieri, R., Arnese, A., Prospero, E., Roncioni, A., Baruffo, L., Ullucci, R., Romano, V., 1996, Lead in human blood from children living in Campania, Italy, *J. Toxicol. Environ. Health*, 47: 311–320.

Fredricks, R.E., Tanaka, K.R., Valentine, W.N., 1960, Zinc in human blood cells: normal values and abnormalities associated with liver disease, *J. Clin. Invest.*, 39: 1651–1656.

Hallen, I.P., Jorhem, L., Lagerkvist, B.J., Oskarsson, A., 1995, Lead and cadmium levels in human milk and blood, *Sci. Total Environ.*, 166: 149–155.

Schwaiger, M., Mueck, K., Benesch, T., Feichtinger, J., Hrnecek, E., Lovranich, E., 2004, Investigation of food contamination since the Chernobyl fallout in Austria, *Appl. Radiat. Isotopes*, 61: 357–360.

Tefas, L., Ossian, A., Ionut, R., Bocsa, H., 1998, The metabolism of some essential elements (copper and zinc) in experimental lead poisoning, In: 5th Internet World Congress on Biomedical Sciences, McMaster University, Canada, http://www.mcmaster.ca/inabis98/occupational/tefas0267/index.html

ORGANOCHLORINE POLLUTION OF MARICULTURE OBJECTS OF THE CRIMEA COASTAL AREA (BLACK SEA)

LUDMILA MALAKHOVA
Institute of Biology of the Southern Seas National Academy of Sciences of Ukraine, Sevastopol, Crimea, Ukraine

VLADIMIR VORONOV
Irkutsk State Technical University, Irkutsk, Russia

Abstract: The main purpose of this work was to determine the concentrations of polychlorinated biphenyls (PCBs) and organochlorine pesticides (OCs) in the Black Sea mussels *Mytilus galloprovincialis* Lam, which are most promising cultivation object. Also mussels, filter feeder and sedentary organisms, were used in order to test water pollution of the Crimea coast (Ukrainian Region of the Black Sea). All samples contained different concentrations of PCBs and the dominant congeners of PCBs were from tetra to hexachlorobiphenyls among these, congeners 153 and 138 were the most representative. Among the OCs the highest concentrations were found for α-HCH and p,p'-DDE and p,p'-DDD that are metabolite of DDT. Results have also been interpreted in terms of geographical distribution and mussels' biological cycle.

Keywords: Polychlorinated biphenyls, organochlorine pesticides, mussels *Mytilus galloprovincialis*, the Black Sea

1. Introduction

The Black Sea is located between the European and Asian continents and linked to the Mediterranean Sea through the Sea of Marmara. It receives pollutants from Ukraine and other Black Sea riparian states, and through rivers from the Eastern European countries, such as Bulgaria, Romania, Austria, etc. as well as former USSR countries. The present environmental problems are due to unmanaged shipping activity, untreated sewage discharge by coastal settlements, dumping of toxic and industrial wastes from its surrounding countries. Though some studies have been done to assess the environmental quality of the Black Sea (Mee, 1992), the aspects on the contamination by toxic organochlorines especially in aquatic organisms are barely commencing in this area. Hence, comprehensive studies are needed to understand the status of pollution and biological effects of these chemicals in the Black Sea.

Mussels are the most promising cultivation object in water areas of the Crimea coast of the Black Sea. Mussels are used world-wide as sentinel organisms to rapidly assess the status of the contamination of the marine environment for a large number of pollutants. They offer the advantage of a wide geographic distribution, facilitating comparison of data, and of integrating chemical pollutants over long periods at the same site (Farrington et al., 1987).

Synthetic organochlorines such as DDTs, PCBs (polychlorinated biphenyls) and HCHs (hexachlorocyclohexanes) are highly resistant to degradation by biological, photochemical or chemical means. They are also liable to bioaccumulation, are toxic and probably hazardous to human and/or environmental health. Most are prone to long-range transport (UNEP, 1996). These compounds are also typically characterized as having low water solubility and high lipid solubility (Carvalho et al., 1994). The orgonochlorines have been associated with significant environmental impact in a wide range of species and at virtually all tropic levels. Many organochlorines have been implicated in a broad range of adverse human health and environmental effects, including impaired reproduction, endocrine disruption, immunodepression and cancer (UNEP, 1996). The

United Nations Environment Programme (UNEP) chose 12 POPs, all chlorine-containing organic compounds, as priority pollutants due to their impact on the human health and environment. Polychlorinated biphenyls and organochlorine insecticides were also included in the list (Euro Chlor, 1998–1999).

In marine organisms, chlorinated compounds uptake occurs directly from the sea and through the food chain. Bioaccumulation depends on the uptake, elimination ability of each organism and the physicochemical properties of the xenobiotic (Walker et al., 1996).

The primary transport routes into marine and coastal environments include atmospheric deposition and surface run-off, the former being by far the greatest albeit dispersed over large areas. Because many organochlorines are relatively volatile, their remobilization and long-distance redistribution through atmospheric pathways often complicates the identification of specific sources. Nevertheless, those (the majority) used in agriculture are also washed off the land into rivers, thence to the sea or directly into the sea via outfalls or run-off. Many organohalogens follow quite complex biogeochemical seas pathways including Black Sea (UNEP/IAEA/FAO, 1992).

2. Materials and methods

2.1. SAMPLING

Four different stations, described in Table 1, were chosen along the Crimea coast in 2005. Sampling scheme is shown in Figure 1. All samplings were performed from January to May 2005.

TABLE 1. Description of coastal sampling stations

N° of station	Sampling location	Properties	Date of collection	Coordinates	Sampling depth (m)
1	Streletskaya Bay (Sevastopol Bay region)	Dockyard, shipyard,	01.26.05	44°35′670 N 33°28′200 E	1.2
		industrial and domestic sewage	04.22.05	44°35′670 N 33°28′200 E	1.2
2	Laspi Bay	Moorings, delphinarium,	01.28.05	44°25′130 N 33°42′570 E	1.2
		waste sewage	05.31.05	44°25′165 N 33°41′481 E	Up to 2
3	Karadag Bay (Gold gate)	Sea reservation	03.31.05	44°54′877 N 35°13′870 E	Up to 3
	Karadag Bay (Kurortnoye village)		03.29.05	44°55′502 N 35°11′301 E	4
	Karadag Bay (Gold gate)		04.27.05	44°54′877 N 35°13′870 E	Up to 3
	Karadag Bay (Kurortnoye village)		04.26.05	44°55′502 N 35°11′301 E	4
4	Tarkhankut Cape, Donuzlav Lake	Moorings, industrial and	02.28.05	45°23′245 N 33°06′934 E	Up to 2
		domestic sewage	05.04.05	45°23′245 N 33°06′934 E	Up to 2

At each site, a large number of mussels of similar size, 3–5 cm shell length were collected and transported in the laboratory, where have been determined stage of the gametogenesis cycle. Mussels were opened with stainless steel knives and the shell lengths were measured. The flesh of mussel was rinsed with distilled water to remove sand and impurities. The flesh of mussel was homogenized using a blender.

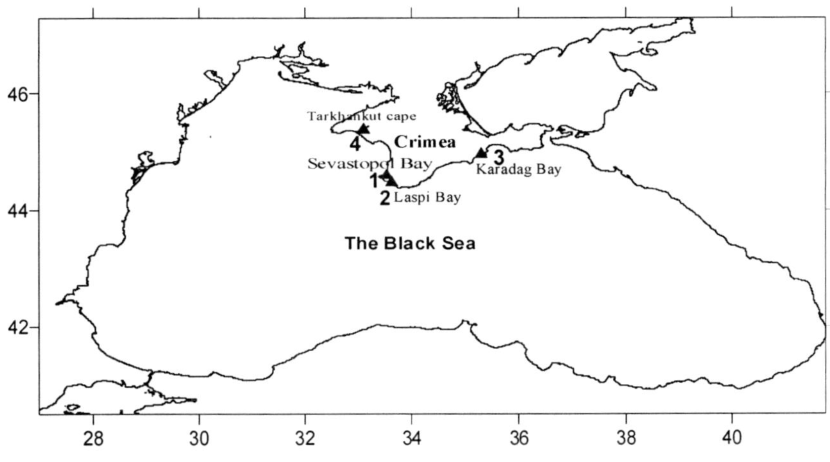

Figure 1. Sampling location

2.2. ANALYTICAL METHODS

Pesticide-grade reagents and solvents were used for the analysis. Pesticide-grade reagents and solvents were used for the analysis. Five to 10 g of sample of mussel was mixed with sodium sulfate and was extracted in a Soxhlet apparatus with *n*-hexane. The lipid content of each sample was gravimetrically determined.

The cleanup was carried out by concentrated sulphuric acid. The organic phase passed through a silica gel column with sodium sulphate anhydrous at the top level. The solvent phase was concentrated and stored at 4°C before analyzing by gas chromatography.

Quantification of organochlorine residues was made on gas chromatograph (Varian 3800) equipped with a ^{63}Ni electron capture detector and moving needle-type injection port. A fused silica capillary column (25 m × 0.25 mm i.d.) coated with CP Sil-8CB (-Si-C_8H_{17}, 5%-phenyl, 96%-dimethyl polysiloxane) having a film thickness of 0.45 gm was used for the determination of PCBs and OCs. Oven operated under the following conditions: 90°C, 1 min hold, ramp to 120 at a rate of 25°C/min, 0.5 min hold, ramp to 290°C at a rate of 5°C/min, 9 min hold with a final hold of 10 min. Injection 1 μl into a split/splitless injector used in splitless mode, with 1 min of injection time. The injector and detector temperatures were kept at 250°C and 300°C. Nitrogen was used as the carrier and the make-up gas.

Ten PCB congeners IUPAC numbers 18, 26, 31, 52, 101, 118, 138, 153, 180 and 183 were quantified using a mixed PCBs standard obtained from Supelco. These PCB congeners were selected because they contain six congeners of the International Council for Exploration of the Sea list: 52, 101, 118, 138, 153 and 180 proposed as pollution indicators (Duinker et al., 1988).

Calibration curve of individual compounds was constructed on three concentration levels.

Detection limits for the method used for individual PCB congeners were: <0.05 ng·g^{-1} (118 and 180), <0.06 ng·g^{-1} (101), <0.07 ng·g^{-1} (52), <0.08 ng·g^{-1} (138), <0.10 ng·g^{-1} (153 and 209), for OCs – 0.1 ng·g^{-1} (α- and γ-HCH), 0.05 ng·g^{-1} (p,p'-DDD, p,p'-DDE) and 1 ng·g^{-1} (Aroclor 1254). Quality insurance criteria have been applied before samples analysis. Regular analyses of reference material of marine biota for PCBs and OCs from the International Atomic Energy Agency (MESL/IAEA-435, IAEA-432) gave satisfactory results.

3. Results

The localization of sampling, hexane extractable organic matter, concentrations of chlorinated pesticides and PCBs in mussel samples are shown in Table 2.

PCB concentrations were between 2.9 and 145 ng g^{-1} WW. These PCB corresponds to 42% of a standard commercial Aroclor 1254, although the patterns change significantly in the environment. It was shown, that the maximal PCB concentration was 115 ng g^{-1} WW for mussels from Streletskaya Bay. The minimum of the PCB contents in mussels was 7 ng·g^{-1} WW for mussels from Karadag Bay. Some decreasing of PCB concentrations was determined for mussels from all sampling locations for the investigated period.

TABLE 2. Average contents hexane extractable organic matter (HEOM, %), concentrations of α- and γ-HCH, p,p'-DDE, p,p'-DDD and PCB (Aroclor 1254) (range and mean values) in mussels of Crimea coast

N° of station	Date of collection	HEOM %	Concentration (ng g^{-1} WW)				
			α- HCH	γ- HCH	p,p'-DDD	p,p'-DDE	PCB Aroclor 1254
1	01.26.05	1.22–1.88 1.52	0.81–4.02 2.22	0.84–4.02 0.51	0.30–0.73 0.60	0.41–1.33 1.01	101–217 145
2	01.28.05	0.23–0.54 0.39	2.31–4.05 3.48	1.01–5.13 3.47	n.d.*	9.91–22.14 16.02	89–115 102
3	03.29.05	0.88–0.95 0.92	10.90–17.00 13.95	n.d.	n.d.	n.d.	8.6–9.2 8.9
3	03.31.05	0.82–1.43 1.13	7.90–9.52 8.95	n.d.	n.d.	n.d.	5.1–14.2 9.7
4	02.28.05	0.97–1.13 1.19	7.58–13.24 10.29	0.10–0.13 0.12	1.02–2.80 1.96	1.26–3.17 2.80	36–78 59
1	04.22.05	1.02–1.25 1.18	10.5–15.00 12.75	n.d.	n.d.	n.d.	68–100 85
2	05.31.05	1.40–1.50 1.44	1.80–2.80 2.30	n.d.	5.50	n.d.	2.9–9.2 6.0
3	04.27.05	0.79–1.64 1.22	7.02–15.60 11.31	n.d.	n.d.	n.d.	3.0–4.3 3.5
4	05.04.05	0.76–1.15 0.97	9.00–16.20 14.50	n.d.	n.d.	n.d.	18–60 39

The concentrations of metabolite DDTs: p,p'-DDE and p,p'-DDD ranging from 0.30 to 22.14 ng g^{-1} WW. The maximal average p,p'-DDE concentration was 16.02 ng g^{-1} WW for mussels from Laspi Bay.

HCH concentrations (α and γ-isomers) were found to be in the range of 0.10–17 ng g^{-1} WW (Table 2). Concentrations of lindane (γ-HCH) and α-HCH were low in samples from Laspi bay (Figure 1). Elevated concentrations in samples from stations 3, 4 and 1 (spring) indicated usage of HCH as a pesticide in these areas. α-Isomer of HCH was dominant at all stations.

In 2005, were analyzed hermaphrodite, females and males of mussel on 1–6 stages of gametogenesis (Table 3).

There was observed differences on PCB between age groups. The young adult mussels (5 months old) had a higher concentration of PCB 26, 52, 101, 138 (Table 4).

TABLE 3. The stages of the Black Sea mussel *M. galloprovincialis* gametogenesis cycle (Pirkova, 1994)

No stage	Name of stage
1	The relative quiescence after spawning
2	The beginning of gametogenesis
3	The active gametogenesis
4	The stage of gametogenesis before the spawning
5	The spawning itself, gonads release sexual products
6	The realignment after spawning

TABLE 4. PCBs concentration levels determined in selected mussel *Mytilus galloprovincialis* collected along the Crimea coast in 2005. Results are concentration $\pm \sigma$ ng g^{-1} wet weight

Compounds	Station 1[b] $\female$[c], 3[d]	Concentrations of PCB $\pm \sigma$[a] ng g^{-1} wet weight				
		Station 1		Martinova Bay (Sevastopol Bay region)		
		$\male$, 4[d]	$\female$, 4	Age of mussels − 5 months	Age of mussels − 1 year	Age of mussels − 2.5 years
PCB 18	−[e]	7 ± 1,1	8 ± 1	−	−	−
PCB26	−	−	−	27 ± 4	11 ± 2	8 ± 1
PCB 31	−	3 ± 0.5	6 ± 0.9	−	−	−
PCB 52	18 ± 3	5 ± 0.7	9 ± 1.4	35 ± 5	29 ± 4	24 ± 4
PCB 101	16 ± 2	2 ± 0.3	5 ± 0.7	15 ± 2	10 ± 1,5	6 ± 1
PCB 151	−	−	−	−	−	−
PCB 153	32 ± 5	5 ± 0,8	10 ± 1.5	13 ± 2	20 ± 3	7 ± 1
PCB 138	27 ± 4	4 + 0,6	8 + 1 2	20 + 3	17 + 2 6	5 ± 0.7
PCB 183	5 ± 0.7	−	−	3 ± 0.5	3 ± 0.5	1 ± 0.2

[a]σ – 1 σ in ng g^{-1} wet weight
[b]Sampling location see in Table 1
[c]Sex: $\male$ − male, − female
[d]Stage of the gametogenesis cycle (see in Table 3)
[e]Not detected

All samples contained different concentrations of PCBs and the dominant congeners of PCBs were from tetra to hexachlorobiphenyls among these, congeners 153 and 138 were the most representative (Table 4).

4. Discussion

Residues of OCs and PCBs were measured in every single station. PCB and OCs concentrations were depending on the location. The highest concentrations measured 22 and 15 ng g^{-1} wet wt. for DDE in mussels from Laspi bay and Karadag bay and 217 ng g^{-1} wet wt. for PCBs corresponded to major coastal cities and industrial sites, such as Sevastopol. The high concentrations of PCBs associated with coastal areas of industrial cities indicated that the cities were a major source of contamination of PCB to the Crimea coastal area (Figure 2). The minimum of the PCB contents was 7 ng·g^{-1} WW (range 3–15 ng·g^{-1} WW) for mussels from recreation area of the Karadag Bay.

Amongst the DDT compounds, p,p'-DDE and p,p'-DDD were measured at concentrations frequently higher than those of p,p'-DDT, suggesting bioaccumulation of DDT from old applications. This suggests that recently these pesticides have not been used in agriculture after their ban. These results contrasted with the results obtained 20 years earlier, in which DDT dominated over DDT metabolites as a result of exposure to freshly applied DDT (Polikarpov et al., 1996) and indicating extensive usage of this insecticide.

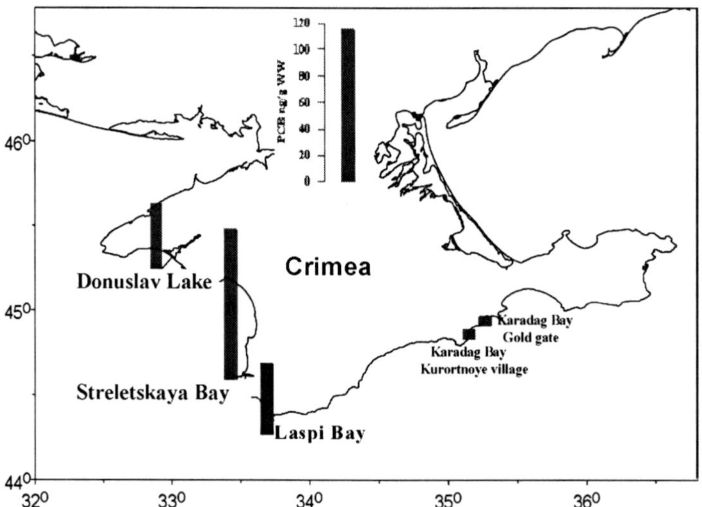

Figure 2. Mean concentrations of PCBs (Aroclor 1254, ng g^{-1} WW) in mussels *Mytilus galloprovincialis* in 2005

Domination of α-isomer of HCH in mussels possible indicate use in this area of a technical mix HCH (which is used as a pesticide) was reported as a mixture of β (c. 9%), α (c. 14%) and α (c. 70%) isomers (Iwata et al., 1995).

The total PCBs concentrations were significantly correlated with the hexane extractable organic matter determination coefficient ($r = 0.54$) of the mussel, but indicating that the measured PCB concentrations cannot be explained only by the affinity of these compounds for the organic matter. It was shown that the PCB concentrations were higher in mussels on four and five stages of gametogenesis (Figure 3).

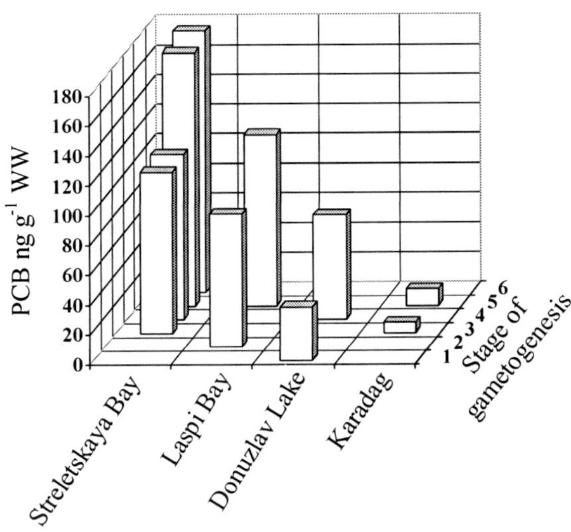

Figure 3. Average of PCB in mussels of 1–6 stage of gametogenesis

5. Conclusion

These results demonstrate the existence of point sources where PCB and OCs concentrations are still high, despite the observed general decrease due to reduction of production and use of these compounds and, hence, subsequent releases into the marine environment. The highest PCBs concentrations correspond to industrial sites, such as Streletskaya Bay. It is shown that the PCB concentration was depended from physiological condition mussel: stage of gametogenesis and contents hexane extractable organic matter.

References

Carvalho, F.P., Fowler, S.W., Readman, J.W., Mee, L.D., 1994, Pesticide residues in tropical coastal lagoons use of 14C labelled compounds to study the cycling and fate of agrochemicals. Int. Symp. Applications of Isotopes and Radiation in Conservation of the Environment, 9–13 March 1992, Karlsruhe, Germany, Vienna, International Atomic Energy Agency, Austria, 637–653.

Duinker, J.C., Schulz, S.E., Petrick, G., 1988, Selection of chlorinated biphenyl congeners for analysis in environment samples., *Mar. Pollut. Bull.*, 19: 19–25.

Euro Chlor (European Chlor-Alkali Industry), 1998–1999, Chlorine Industry Review.

Farrington, J.W., Davis, A.C., Tripp, B.W., Phelps, D.K., Galloway, W.B., 1987, 'Mussel Watch' — measurements of chemical pollutants in bivalves as one indicator of coastal environment quality. In: New approaches to monitoring aquatic ecosystems, T.P. Boyle (Ed). ASTM STP 940. American Society for Testing and Materials, Philadelphia, 125–139.

Iwata, H., Tanabe, S., Ueda, K., Tatsukawa, R., 1995, Persistents organochlorine residues in air, water, sediments, and soils from the Lake Baikal region, Russia, *Environ. Sci. Technol.*, 29: 792–801.

Mee, L.D., 1992, The Black Sea in crisis, a need for concerted international action, *Ambio*, 21: 278–286.

Pirkova, A.B., 1994, The reproduction of mussel *Mytilus galloprovincialis* Lam. and an elements of its cultivation: Abstract of PhD thesis. Sevastopol: IBSS. 25 p. (Russian).

Polikarpov, G.G., Zherko, N.V., 1996, Ecological aspects studying of organochlorine pollution of Black sea, *Ecol. Sea*, 45: 93–101 (Russian).

UNEP, 1996, State of the marine and coastal environment in the Mediterranean Region. Mediterranean Action Plan Technical Report Series 100, UNEP. Athens, 138 pp.

UNEP/IAEA/FAO, 1992, Organohalogen compounds in marine environment. A review, Mediterranean Action Plan Technical Report Series 70, UNEP. Athens, 49 pp.

Walker, C.H., Hopkin, S.P., Sibly, R.M., Peakall, D.B., 1996, Principles of ecotoxicology, Taylor & Francis, London.

World-wide and regional intercomparison for the determination of organochlorine compounds and petroleum hydrocarbons in mussel tissue, IAEA-432, 2004, Vienna, Austria, 118 pp.

World-wide and regional intercomparison for the determination of organochlorine compounds and petroleum hydrocarbons in tuna gomogenate sample IAEA-435, 2006, Vienna. Austria, 126 pp.

AFLATOXIN CONTAMINATION RISK: BIOACTIVE NATURAL COMPOUNDS FOR ANIMAL HEALTH AND HEALTHY FOOD

DORIANA TEDESCO, CHIARA BARBIERI, STEFANO LUGANO, LAURA GARAVAGLIA
Department of Veterinary Science and Technologies for Food Safety, Milan, Italy

Abstract: Mycotoxin contamination is a worldwide problem and significant economic losses are associated with their impact on human health, animal productivity and both domestic and international trade. The FAO has estimated that up to 25% of the world's food crops and a higher percentage of the world's animal feedstuffs are significantly contaminated by mycotoxins and represent a human safety risk. Aflatoxin B_1 (AFB_1), a potentially lethal metabolite, is a known human carcinogen. AFB_1 is toxic to the liver, immunosuppressant, hepatocarcinogenic, teratogenic, and mutagenic. Aflatoxin carryover to animal products poses a health and economic liability and drives the demand for a method to prevent aflatoxin contamination of feedstuffs and/or animal food products such as milk, eggs, and meat. One of the innovative methods to overcome the toxic and carcinogenic effects of AFB_1 is to enhance its metabolism by natural substances reportedly possessing hepatoprotective effects. In this study we reported the effect of silymarin, a potent antihepatotoxic agent used as a hepatoprotector in man, on reducing aflatoxin M_1 (AFM_1) excretion in cows' milk and the effects of a silymarin in reducing the toxic effects of AFB_1 in broiler chickens.

Keywords: Aflatoxin, animal contamination, hepatoprotector, milk excretion, chicken

1. Introduction

Mycotoxins are naturally occurring secondary toxic fungal metabolites that contaminate various agricultural products, either pre- or post- harvest. Mycotoxin contamination of foods and animal feedstuff is a worldwide problem. No region of the world escapes the problem of mycotoxins, however, in certain geographical areas of the world, some mycotoxins are produced more readily than others (Lawlor and Lynch, 2005). The FAO has estimated that up to 25% of the world's food crops and a higher percentage of the world's animal feedstuffs are significantly contaminated by mycotoxins and represent a human safety risk (FAO, 2004). The Rapid Alert System for Food and Feed (RASFF), a network involving the Member States, the European Commission and the European Food Safety Authority (EFSA) alerts when a residue of potential concern has been detected in food of domestic or imported origin. Mycotoxins are the hazard category with the highest number of notifications of which 802 concerning aflatoxins (RASFF, 2007). Human exposure to mycotoxins may result from consumption of plant derived foods that are contaminated with toxins and the carryover of mycotoxins and their metabolites into animal products such as milk, meat, visceral organ and eggs or exposure to air and dust containing toxins (Jarvis, 2002). Aflatoxins are toxic to the liver, immunosuppressant, hepatocarcinogenic, teratogenic, and mutagenic. The most important genera of mycotoxigenic fungi are *Aspergillus* and *Fusarium*. These are prevalent in areas located in tropical regions between latitudes 40°N and 40°S of the equator. Both intrinsic and extrinsic factors influence fungal growth and mycotoxin production. The intrinsic factors include water activity (aw = 0.78), pH, and redox potential whereas extrinsic factors are relative humidity, temperature and availability of oxygen. *Aspergillus* and *Fusarium* grow on certain foodstuffs, most commonly groundnuts, dried fruit, tree nuts, spices, cereals. In these products aflatoxins B_1 (AFB_1), B_2, G_1, G_2 are found. Aflatoxins are efficiently absorbed in the intestinal tract, probably

B. Faye and Y. Sinyavskiy (eds.), *Impact of Pollution on Animal Products.*
© Springer Science + Business Media B.V. 2008

by passive diffusion (Hsieh and Wong, 1994). Biotransformation achieves predominantly by cytochrome P-450 (CYP-450) enzyme systems in the liver. The main direct products of phase 1 metabolism are AFM_1 (by CYP-450 1A2), AFQ_1 (by CYP-450 3A4) and AFB_1-8,9-epoxide (by CYP-450 3A4) (Kamdem et al., 2006). AFM_1 is secreted into milk when lactating animals are fed with AFB_1 contaminated diets (Patterson et al., 1980). Aflatoxicol (AFL) can also be generated by cytosolic reductases (Eaton, 1994). However, the carcinogenic effect of AFB_1 is regarded to occur via production of the highly reactive AFB_1-8,9-epoxide (AFBO) which binds to DNA to form 8,9-dihydro-8-(N7guanyl)-9-hydroxy-AFB_1 (AFB_1 N7-Gua) adduct (Sharma and Farmer, 2004; Klein et al., 2002) and a lysine adduct in serum albumin (Sabbioni et al., 1987). AFB_1-8,9-epoxide is further metabolized by phase 2 enzymes, glutathione-S-transferases (GSTs), to AFB–NAC, a detoxification metabolite excreted in urine (Scholl et al., 1997; Wang et al., 1999). The toxicity may ensue through the generation of intracellular reactive oxygen species (ROS) like superoxide anion, hydroxyl radical and hydrogen peroxide (H_2O_2) during the metabolic processing of AFB_1 by CYP-450 in the liver (Towner et al., 2003; Sohn et al., 2003).

AFB_1 is considered to be the most potent carcinogen among all aflatoxins and is classified as a group I carcinogen in humans (IARC, 1987). AFB_1 consumption contributes significantly to the high incidence of hepatocellular carcinoma (HCC), the sixth most prevalent cancer worldwide with a higher incidence rate within developing countries (Wild and Hall, 1996; Parkin et al., 2005) especially in individuals infected with hepatitis B or C virus (Henry et al., 1999). Moreover AFB_1 could also interact with HIV/AIDS. Preliminary evidence suggests that there may be an interaction between chronic aflatoxin exposure and malnutrition, immunosuppression, impaired growth, and diseases such as malaria and HIV/AIDS (Miller et al., 1994). The chronic incidence of aflatoxin in diets is evident from the presence of AFM_1 in human breast milk (Polychronaki et al., 2007). Aflatoxin exposure in children is also associated with child stunting and child neurological impairment increases markedly following weaning and was associated with reduced growth (Gong et al., 2003). The manifestation of chronic or acute toxicosis as well as carcinogenicity depends on the dose, duration of exposure, and rate of metabolism to less toxic metabolites. Animals have a species-specific sensitivity to aflatoxins (Zaghini et al., 2005). Their presence in feed, even in very low dietary levels (parts per billion), can cause decreases in growth or hepatic toxemia in commercial animals.[4] As well as affecting productivity, mycotoxins can also migrate into animal products and enter the human food chain. Foods derived from animals which have ingested aflatoxin may contain toxic metabolites and the potential danger to man is enhanced because the animals may show no outward signs of disease. The most dangerous example of this is AFB_1 contamination in feed for dairy cow, being converted by the cow into a related toxin AFM_1 ending up in milk and dairy products.

On a worldwide basis, particularly some industrialized countries, had mycotoxin regulations for food and/or feed in 2003, an increase of approximately 30% compared to 1995 (FAO, 2004). Few countries regulate AFB_1 in feedstuffs for dairy cattle. Limiting AFB_1 in animal feeds is the most effective means of controlling aflatoxin M_1 in milk. A limit of 5 µg AFB_1/kg feed for dairy cow and a limit of 20 µg AFB_1/kg in feed for cattle, sheep, goats, swine and poultry are applied in the EU countries (EU, Commission Directive 2003/100/EC). This limit is applied by countries in the European Free Trade

---

[4] Bastianelli D., Le Bas C., 2000. Evaluating the Role of Animal Feed in Food Safety: Perspectives for Action. Proc. of the International Workshop, CIRAD-FAO: "Food Safety Management in Developing Countries". Montpellier (France), 11–13 December 2000.

Association (EFTA), in many of the candidate EU countries and sporadically outside Europe. A limit of 20 μg AFB_1/kg feed for dairy animals and a limit of 100 μg AFB_1/kg intended for breeding beef cattle, breeding swine, or mature poultry is applied in the United States (FDA Compliance Policy Guide 7126.33), Africa and Latin America (FAO, 2004). Regulations for AFM_1 existed in 60 countries at the end of 2003, a more than threefold increase as compared to 1995 (FAO, 2004). EU, EFTA, candidate EU countries and some other countries in Africa, Asia and Latin America (FAO, 2004), apply a maximum level of 0.05 μg AFM_1/kg in milk and a maximum level of 0.025 μg AFM_1/kg in infant formulae (EU, Reg. 1881/2006). A limit of 0.5 μg AFM_1/kg in milk is applied in the United States (FDA Compliance Policy Guide 7106.10), several Asian, European countries and in Latin America, where it is also established as a harmonized MERCOSUR (a trading block consisting of Argentina, Brazil, Paraguay and Uruguay) limit (FAO, 2004).

To detect AFB_1 and AFM_1 contamination, a commercial enzyme-linked immunosorbent assay (ELISA) kit, a simple, cheap and rapid method, is adopted as in-house screening method for individual mycotoxins. ELISA may produce false-positive results, however, due to antigen-antibody cross-reactions, so further confirmation using HPLC or GLC is sometimes necessary (Scudamore, 2005).

As reported in the review by Jouany (2007), good agricultural practices during both pre- and post-harvest minimize the risks of contamination by mycotoxins. However the contamination can occur and AFB_1 is a common contaminant in certain animal feedstuff. Methods for reducing levels of aflatoxins include physical, chemical and biological treatments. The addition of the non-nutritionally adsorbents in the diet, synthetic cation or anion exchange zeolite, bentonite, hydrated sodium calcium aluminosilicate (HSCAS), yeast cell wall preparations and *Lactobacillus rhamnosus* strain GG (Gratz et al., 2007), bind mycotoxins in the gastrointestinal tract and reduce their bioavailability. One of the innovative methods to overcome the toxic and carcinogenic effects of AFB_1 is to enhance AFB_1 metabolism toward its detoxification in humans or animals. Recently natural substances that can prevent AFB_1 hepato-carcinogenesis and hepatotoxicity would be helpful to human and animal health with minimal cost in foods and feed. Coumarin, a natural antioxidant contained in a variety of plant, may affect AFB_1 metabolism toward the enhancement of detoxification through the suppression of CYP-450 enzymes in the liver and intestine, and through the enhancement of GSTs toward AFB_1 in the intestine (Tulayakul et al., 2007). *Laurencia obtusa* extract and the *Caulerpa prolifera* extract, marine algae collected from the Egyptian Coast of the Red Sea Sinai, could stimulate the antioxidant defence system and hepatic cell regeneration (Abdel-Wahhab et al., 2006). *Aquilegia vulgaris*, a perennial herb indigenous in central and southern Europe and in Asia, attenuate AFB_1-induced hepatic injury as evidenced by the inhibition of lipid peroxidation, preventing reduced glutathione depletion and the decrease in transaminases leakage to serum (Jodynis-Liebert et al., 2006). Lupeol, isolated from the medicinal plant *Crataeva nurvala* Buch-Ham (Capparidaceae), has enhanced the antioxidant status and produced substantial protection against AFB_1-induced liver injury. It showed a trend similar to that of silymarin, a known hepato-protective agent in protecting liver from AFB_1-induced toxicity (Preetha et al., 2006). Silibin, a major constituent of silymarin, evidenced the inhibition of hepatic CYP-450 detoxification system and the protection against free radicals generated in the phase 1 metabolism (Baer-Dubowska et al., 1998; Rastogi et al., 2000). Because all AFB metabolizing enzymes are expressed in the liver at individually variable levels, targeted chemoprotection by CYP-450 inhibition represents a valid hypothesis. Considering the prominent contribution of CYP-450 3A4 to AFBO formation, the inhibition of CYP-450 3A4 should be much more effective (Kensler et al., 2003). Thus, silymarin pre-treatment and intervention could significantly blocked phase 1 metabolism and metabolic activation of AFB_1 and could greatly induced phase 2 detoxifying enzymes by increase the activity of GSTs.

Briefly, we reported some trials performed to evaluate the hepatoprotective activity of silymarin to counteract AFB_1 intoxication in dairy cow and poultry (Tedesco et al., 2003, 2004).

2. Materials and Methods

2.1. TRIAL 1

An organic dairy farm with a high level of AFM_1 (87 ng/l) in bulk milk yield, was used. Twenty lactating Italian Friesian dairy cows were selected according to their milk production, days in lactation and parity. The first treatment lasted 9 days in which 10 cows were treated with 10 g/day of silymarin (76% pure extract). The extract was mixed with water and the solution was immediately administered by oral drench. The silymarin mixture used in this trial, previously analyzed by HPLC was composed of: silybin 49.8%, isosilybin 14.7%, silydianin 20.1%, silycristin 6.1% and taxifolin 4.5%. AFB_1 in feed samples were determined by HPLC and I2 post-column derivatization (Paulsch et al., 1988). Quantitative detection of AFM_1 in milk samples was tested in triplicate using ELISA. (Riedel-de Haën, Sigma-Aldrich, Laborchemikalien GmbH, Seelze, Germany).

2.2. TRIAL 2

A total of 21 fourteen-day-old male commercial broilers (377 ± 34 g body weight) were randomly allotted into three groups of seven animals each. Dietary treatment lasted for 35 days and consisted of: AFB_1 at 0.8 mg/kg of feed (group B1), AFB_1 at of 0.8 mg/kg of feed plus silymarin at 600 mg/kg of BW (group B1 + Sil), control group (group C) was fed on a basal diet alone. Chicks were weighed individually every week. Feed intake was recorded daily in the last 2 weeks of the experimental period. Morbidity and mortality were recorded in each group. Before slaughter, blood samples were collected from the brachial vein, centrifuged and sera were stored at −20°C. Samples were submitted for analysis in order to measure total protein, albumin, globulin, glucose, urea, total bilirubin, direct bilirubin, indirect bilirubin, aspartate amino transferase (AST), γ glutamyl transferase (GGT), alanine amino transferase (ALT), calcium, phosphorus, using an automatic analyser (Roche BM Hitachi 911, Hitachi Medical Systems, Tarrytown, NY). Liver weight was recorded for each animal and two animals for each group were randomly selected for histological studies on liver tissue.

3. Results

3.1. TRIAL 1

Values of AFM_1 concentrations in milk are presented in Figure 1. In order to account for individual variability, AFM_1 values of the first two samples (day 5 and 0) were used as covariates. For the whole period of treatment a lower emission of AFM_1 in treated animals was observed. On day 3 AFM_1 value was significantly lower in the treated

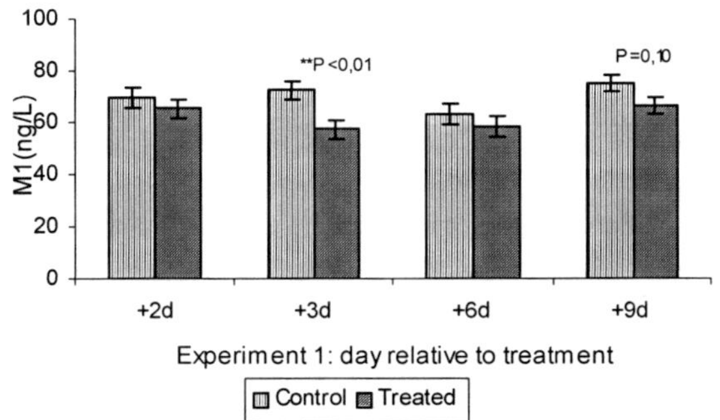

Figure 1. Mean AFM$_1$ concentration (ng/l) in milk (first trial: treatment = silymarin, n = 10). **P < 0.01

group (57.5 vs. 72.7 ng/l, p < 0.01) and there was a tendency for treatment effect on day 9 (66.4 vs 75.1 ng/l, p = 0.10).

3.2. TRIAL 2

No morbidity or mortality due to aflatoxin ingestion was recorded in this study. The average BW of each experimental group is shown in Figure 2. In contrast with the significant effect on growth performance induced by AFB$_1$ alone, there were no differences in BW between control and B1 + Sil groups. The mean feed intake in the last 2 wk of experimental period was lower in AFB$_1$-treated birds (P < 0.05) compared with the other groups. Histological sections from the livers of AFB1-treated birds showed multifocal portal infiltration composed of mononucleates, granulocytes, and eosinophiles diffused in the parenchyma, especially at the portal areas, and necrosis in zone 1. These changes were less severe in the birds receiving AFB$_1$ plus silymarin. Control birds presented modest vascular congestion and minimal focal infiltration. The serum activity of alanine amino transferase (ALT), a marker of liver injury, was lower (P < 0.05) in AFB$_1$-treated birds. In birds receiving AFB$_1$ plus silymarin, there was no difference in ALT serum activity compared with controls.

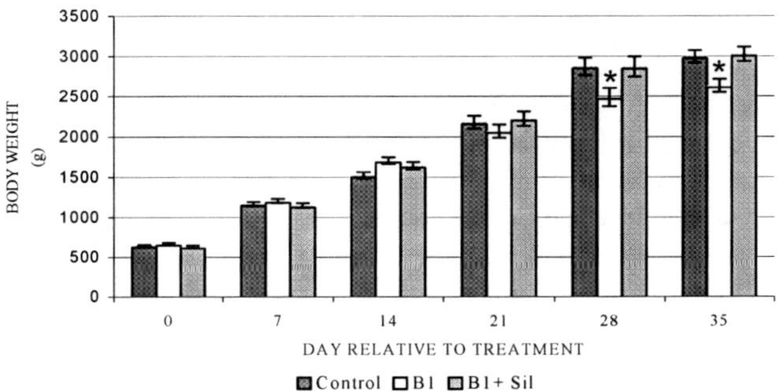

Figure 2. Effect of AFB$_1$ and silymarin on body weight. Control = basal diet alone, B1 = aflatoxin B1 0.08 mg/kg feed, B1 + Sil = aflatoxin B1 0.08 mg/kg plus silymarin 600 mg/kg body weight. Group marked with asterisk differ significantly from the others (*P < 0.05)

4. Discussion

4.1. TRIAL 1

Although silymarin does not completely inhibit AFM_1 emission, the results of this study show that these compounds can contribute to reduce milk contamination. Even though aflatoxin M1 may be considered a less dangerous genotoxic substance than AFB_1, it is still necessary to prevent its presence in milk and milk products intended for human consumption, in particular for young children, who are very sensitive to its effect.

4.2. TRIAL 2

The effect of AFB_1 on performance was evident after 3 wk of treatment. Despite this evident effect of AFB_1 intoxication, there was no difference in performance when the birds received AFB_1 plus silymarin. These results suggest that treatment with silymarin can be effective in counteracting the negative effects of AFB_1 intoxication on feed intake and BW in growing broilers. At necropsy, our observations demonstrate the ability of silymarin to counteract the action of AFB_1, reducing hepatic histopathological changes induced by the toxin and preventing the changes in ALT activity due to a hepatotoxic effect of AFB_1. Silymarin could inhibit the cytochrome CYP-450 3A4 and consequently inhibit the production of the highly reactive AFBO and the formation of DNA adducts and can influence enzyme systems associated with GSTs metabolism to improve safety and quality of poultry products.

5. Conclusion

Mycotoxins are a food safety risk globally and are generally orders of magnitude lower in developed than in developing regions. Furthermore, the consumers are confronted with a diet that contains a low level of toxin and in many cases there may be other toxins present. Thus, a complimentary toxicity mechanism of action occurs (Riley, 1998). Implicit with these conclusions are the existence of syndromes of apparently unknown aetiology and epidemiology that may involve mycotoxins and the difficulty of establishing "no effect" levels for mycotoxins. Following AFB_1 intake, the most recent approach for the prevention of mycotoxicosis concern the use of active principles isolated from plants reportedly possessing hepatoprotective effects. These compounds enhance AFB_1 metabolism toward inhibition of phase I mediated by cytochrome CYP-450 3A4 and induce phase 2 detoxifying enzymes by increase the activity of glutathione peroxidase, GSTs and glutathione reductase. These findings provide a basis for further studies on the relationship between natural compounds and reduced risk of AFB_1 and its metabolites that occur in animal derivative foods.

References

Abdel-Wahhab, M.A., Ahmed, H.H., Hagazi, M.M., 2006, Prevention of aflatoxin B1-initiated hepatotoxicity in rat by marine algae extracts, *J. Appl. Toxicol.*, 26: 229–238

Baer-Dubowska, W., Szafer, H., Krajkakuzniak, V., 1998, Inhibition of murine hepatic cytochrome P450 activities by natural and synthetic phenolic compounds, *Xenobiotica*, 28: 735–743

Eaton, D.L., Ramsdell, H.S., Neal, G.E., 1994, Biotransformation of aflatoxins, p. 45–71. In: The toxicology of aflatoxins: human health, veterinary and agricultural significance, D.L. Eaton and J. Groopman (Eds). Academic, New York, 45–71

European Union (EU), Commission Directive 2003/100/EC of 31 October 2003 amending Annex I to Directive 2002/32/EC of the European Parliament and of the Council on undesirable substances in animal feed, *Off. J. Eur. Commun. 2003*, L 46, 33–37

European Union (EU), Commission Regulation 1881/2006 of 19 December 2006 amending Commission Regulation 466/2001 setting maximum levels for certain contaminants in foodstuffs, *Off. J. Eur. Commun. 2006*, L 364/5

Food and Agriculture Organization of the United Nations (FAO), 2004, FAO Food and Nutrition Paper 81-Worldwide Regulations for Mycotoxins in Food and Feed in 2003. Rome, Italy

Food and Drug Administration (FDA) Sec. 527.400 Whole Milk, Low Fat Milk, Skim Milk – Aflatoxin M1 (CPG 7106.10)

Food and Drug Administration (FDA) Sec. 683.100 Action Levels for Aflatoxins in Animal Feeds (CPG 7126.33)

Gong, Y.Y., Egal, S., Hounsa, A., Turner, P.C., Hall, A.J., Cardwell, K.F., Wild, C.P., 2003, Determinants of Aflatoxin Exposure in Young Children from Benin and Togo, West Africa: The Critical Role of Weaning, *Int. J. Epidemiol.*, 32(4): 556–562

Gratz, S., Wu, Q.K., El-Nezami, H., Juvonen, R.O., Mykkänen, H., Turner, P.C., 2007, Lactobacillus rhamnosus Strain GG Reduces Aflatoxin B1 Transport, Metabolism, and Toxicity in Caco-2 Cells. *Appl. Environ. Microbiol.*, 73, 3958–3964

Henry, S.H., Bosch, F.X., Troxell, T.C., Bolger, P.M., 1999, Reducing liver cancer – global control of aflatoxin. Science, 286: 2453–2454

Hsieh, D., Wong, J.J., 1994. Pharmacokinetics and excretion of aflatoxins. In: The toxicology of aflatoxins: human health, veterinary and agricultural significance, D.L. Eaton and J. Groopman (Eds). Academic, New York, 73–88

International Agency for Research on Cancer-World Health Organization (IARC-WHO), 1987, Monograph on the Evaluation of Carcinogenic Risks in Humans, Suppl. 7

Jarvis, B.B., 2002, Chemistry and toxicology of molds isolated from water-damaged buildings, Mycotoxins and Food Safety, *Adv. Exp. Med. Biol.*, 504: 43–52

Jodynis-Liebert, J., Matławska, I., Bylka, W., Murias, M., 2006, Protective effect of Aquilegia vulgaris L. on Aflatoxin B1-induced hepatic damage in rats, *Environ. Toxicol. Pharmacol.*, 22: 58–63

Jouany, J.P., 2007, Methods for preventing, decontaminating and minimizing the toxicity of mycotoxins in feed, *Anim. Feed Sci. Technol.*, 137(3/4): 342–362

Kamdem, L.K., Meineke, I., Goedtel-Armbrust, U., Brockmoeller, J., Wojnowski, L., 2006, Dominant contribution of P450 3A4 to the hepatic carcinogenic activation of aflatoxin B1, *Chem. Res. Toxicol.*, 19(4): 577–586

Kensler, T.W., Qian, G.S., Chen, J.G., Groopman, J.D., 2003, Translational strategies for cancer prevention in liver, *Nat. Rev. Cancer*, 3: 321–329

Klein, P.J., Van Vleet, T.R., Hall, J.O., Coulombe, R.A. Jr., 2002, Biochemical factors underlying the age-related sensitivity of turkeys to aflatoxin B(1), *Comp. Biochem. Physiol.*, C 132: 193–201

Lawlorn, P.G., Lynchn, P.B., 2005, Mycotoxin management, *Afr. Farm. Food Process.*, 46: 12–13

Miller, D.M., Wilson, D.M., 1994, Veterinary Disease Related to Aflatoxins. In: The toxicology of aflatoxins: human health, veterinary and agricultural significance, D.L. Eaton and J.D. Groopman (Eds). Academic, San Diego, CA, 347–364

Parkin, D.M., Bray, F., Ferlay, J., Pisani, P., 2005, Global cancer statistics, 2002, *CA Cancer J. Clin.*, 55(2): 74–108

Patterson, D.S.P., Glancy, E.M., Roberts, B.A., 1980, The 'carry over' of aflatoxin M1 into the milk of cows fed rations containing a low concentration of aflatoxin B1, *Food Cosmet. Toxicol.*, 18; 35–37

Paulsch, W.E., Sizoo, E.A., van Egmond, H.P., 1988, Liquid chromatographic determination of aflatoxins in feedstuffs containing citrus pulp, *J. Assoc. Off. Anal. Chem.*, 71: 957–961

Polychronaki, N., West, R.M., Turner, P.C., Amra, H., Abdel-Wahhab, M., Mykkaenen, H., El-Nezami, H., 2007, A longitudinal assessment of aflatoxin M1 excretion in breast milk of selected Egyptian mothers, *Food Chem. Toxicol.*, 45(7): 1210–1215

Preetha, S.P., Kanniappan, M., Selvakumar, E., Nagaraj, M., Varalakshmi, P., 2006, Lupeol ameliorates aflatoxin B1-induced peroxidative hepatic damage in rats, *Comp. Biochem. Physiol.*, Part C 143: 333–339

Rapid Alert System for Food and Feed (RASFF), 2007, Annual Report 2006. Office for Official Publications of the European Communities, 22–23

Rastogi, R., Srivastava, A.K., Srivastava, M., Rastogi, K., 2000, Hepatocurative effect of picroliv and silymarin against aflatoxin B1 induced hepatotoxicity in rats, *Planta Med.*, 66: 709–713

Riley, R.T., 1998, Mechanistic interactions of mycotoxins: theoretical considerations. In: Mycotoxins in agriculture and food safety, K.K. Sinha and D. Bhatnagar (Eds). Marcel Dekker, New York, 227–253

Sabbioni, G., Skipper, P.L., Buchi, G., Tannenbaum, S.R., 1987, Isolation and characterization of the major serum albumin adduct formed by aflatoxin B1 in vivo in rats, *Carcinogenesis*, 8: 819–824

Scholl, P.F., Musser, S.M., Groopman, J.D., 1997, Synthesis and characterization of aflatoxin B1 mercapturic acids and their identification in rat urine, *Chem. Res. Toxicol.*, 10: 1144–1151

Scudamore, K.A., 2005, Principles and applications of mycotoxin analysis. In: The mycotoxin blue book, D. Diaz (Ed). Nottingham University Press, Nottingham, UK, 157–185

Sharma, R.A., Farmer, P.B., 2004, Biological relevance of adduct detection to the chemoprevention of cancer, *Clin. Cancer Res.*, 10: 4901–4912

Sohn, S.K., Kim, J.G., Chae, Y.S., Kim, D.H., Lee, N.Y., Suh, J.S., Lee, K.B., 2003, Large-volume leukapheresis using femoral venous access for harvesting peripheral blood stem cells with the Fenwal CS 3000 Plus from normal healthy donors: Predictors of CD34+ cell yield and collection efficiency, *Journal of Physical Apheresis*, 18: 10–15

Tedesco, D., Tameni, M., Steidler, S., Galletti, S., Di Pierro, F., 2003. Effect of silymarin and its phospholipid complex against AFM1 excretion in an organic dairy herd, *Milchwissenschaft*, 58: 416–419

Tedesco, D., Steidler, S., Galletti, S., Tameni, M., Sonzogni, O., Ravarotto, L., 2004, Efficacy of silymarin-phospholipid complex in reducing the toxicity of aflatoxin B1 in broiler chicks, *Poult. Sci.*, 83(11): 1839–1843

Towner, R.A., Qian, S.Y., Kadiiska, M.B., Mason, R.P., 2003, *In vivo* identification of aflatoxin-induced free radicals in rat bile, *Free Radic. Biol. Med.*, 35: 1330–1340

Tulayakul, P., Dong, K.S., Li, J.Y., Manabe, N., Kumagai, S., 2007, The effect of feeding piglets with the diet containing green tea extracts or coumarin on in vitro metabolism of aflatoxin B1 by their tissues, *Toxicol*, 50: 339–348

Wang, J.S., Groopman, J.D., 1999, DNA damage by mycotoxins, *Mutat. Res.*, 424: 167–181

Wild, C.P., Hall, A.J., 1996, Epidemiology of mycotoxinrelated disease, In: The mycota VI. Human and animal relationships, D.H. Howard and J.D. Miller (Eds). Springer Verlag, Berlin, Germany, VI, 213–227

Zaghini, A., Martelli, G., Roncada, P., Simioli, M., Rizzi, L., 2005, Mannanoligosaccharides and aflatoxin B1 in feed for laying hens: effects on egg quality, aflatoxins B1 and M1 residues in eggs, and aflatoxin B1 levels in liver, *Poult. Sci.*, 84(6): 825–832

IMPACT OF SOME MICROELEMENTS ON ANIMAL HEALTH AND PERFORMANCES IN AZERBAIJAN

ELSHAD A. MAMMADOV, ISFENDIYAR J. ALVERDIYEV
Azerbaijan Technological University, Ganja, Azerbaijan

Abstract: Some microelements carry out important biological functions in vital processes and have an influence on the growth and development of animals. However it has been shown that in high concentration these microelements can have a negative influence. Therefore high concentration of these chemical compounds allowed in food products must not exceed a certain limit.

Keyword: Microelements, contamination, dairy efficiency

1. Introduction

Now plenty of materials about the distribution of microelements and their biological role in animals are available (McDowell, 1992). The microelements are in the same time necessary for human and animal metabolism and enzymatic balance for healthy population, and could be a risk of toxicity in case of high pollution of the environment. This problem is complex and has great value. The environment plays an essential role in food safety especially animal products as the main source of microelements in food from animal origin being the environment (soil, water, plant). Environmental contamination from previous years is growing in frightening size, but at the same time the number of people who are concerned with the condition of the environment protection is fortunately also growing. The present paper will contain two parts: (i) a study of the beneficial effect of microelements as supplement in the buffalo diet, and (ii) the negative effect of microelement as part of pollutants in the environment in Azerbaijan.

2. Materials and Methods

2.1. POSITIVE EFFECT OF MICROELEMENTS

A study on the influence of a mix of microelement salts (cobalt, zinc, iodine and manganese) on the protein exchange and dairy efficiency of water buffalos was carried out on 6 lactating females at buffalo breeding farms. Control and experimental buffalos, selected by a principle of analogues, were of average fatness, had a live weight between 375 and 422 kg, were between the age of 8–10 years, with a daily average milk yield from 2.1 up to 6.7 kg, and fat contents of milk – 6.2–10%. Experiments were carried out during different seasons of year: in the spring, in the summer, in the autumn and in the winter. Each series of seasonal experiments consisted of three periods: preparatory, control and treated. The end of the preparatory and the beginning of the control period lasted 10 days. Animals were divided into two groups on three heads in each group. The first served as the control group, and the second was the experimental group. The first day of the control period to the end of the treated period lasted for 30 days. Each animal received separately, strictly by weight, an identical basal diet which was satisfied the food requirements of the animals. At the same time the exact weight accounting milk yield with measured fat content was carried out.

B. Faye and Y. Sinyavskiy (eds.), *Impact of Pollution on Animal Products.*
© Springer Science + Business Media B.V. 2008

During the experimental period the buffalos from control group continued to receive only the basal diet. The animals of the supplemented group for 30 days, received a mix of salts of cobalt chloride in amount of 0.1 mg/kg, manganese sulphate – 0.3 mg/kg, zinc sulphate – 0.3 mg/kg and potassium iodide – 0.4 mg/kg of live weight. This mineral supplementation was mixed with the basal diet.

As the number of animals in each group was only 3, no statistical test was achieved. Only the changes in dairy production and milk composition were reported.

2.2. NEGATIVE EFFECTS OF MICROELEMENTS AS POLLUTANTS

The presence of microelements in polluting environment was evaluated in the area of the western region of Azerbaijan, especially heavy metals linked to the industrial wastes.

3. Results and Discussion

3.1. IMPACT OF MICROELEMENT SUPPLEMENTATION

Data analysis showed the certain positive influence of a mix of microelements on dairy efficiency of lactating buffalos. At spring, the daily average yield of milk increased to 1.8 litres, fat content of milk increased by 0.2% and the protein contents went up approximately 0.1%. The absolute daily average amount of fat and protein in milk of the experimental buffalos increased accordingly on 149 and 74g, in comparison to the control animals (Table 1).

TABLE 1. Changes in milk yield, dairy fat and protein in milk buffalos control and experimental groups of a spring series of experiences (the daily average data)

Buffalos group	Number of animals	Yield of milk in l	Milk composition			
			Fat		Protein	
			In %	Total in g	In %	Total in g
The control period						
Control	3	4.7	6.2	291.4	3.73	175.3
Supplemented	3	6.0	6.5	390.0	3.94	236.4
Day 10 of supplementation						
Control	3	4.9	6.6	323.4	3.58	175.4
Supplemented	3	6.1	6.9	420.9	3.70	225.7
Day 20 of supplementation						
Control	3	4.7	6.7	314.9	3.70	173.9
Supplemented	3	7.8	7.2	561.6	3.96	308.9

In summer, the milk yield reduced to 0.6 l, but fat and protein content of milk increased to 0.90% and 0.81% respectively in comparison with the control buffalos (Table 2).

At autumn 30-day's feeding (additional forage) lactating buffalos, the mix of microelements increased the daily average yield of milk by 0.7–1.1 l and consequently all lactating buffalos had milk with an average of 636 g fat and 373 g protein, more than buffalos in the control group (Table 3).

TABLE 2. Changes in milk yields, dairy fat and milk protein content in buffalos control and experimental groups of a summer series of experiments (daily average data)

Buffalos Group	Number of animals	Yield of milk in l	Milk composition			
			Fat		Protein	
			In %	Total in g	In %	Total in g
At 10-fold reception of microelements						
Control	3	2.5	9.7	242.5	5.21	130.0
Supplemented	3	3.1	9.5	294.5	4.83	149.7
At 20-fold reception of microelements						
Control	3	2.4	9.8	235.2	4.04	97.0
Supplemented	3	2.8	9.9	274.4	4.63	129.6
At 30-fold reception of microelements						
Control	3	2.5	9.6	247.5	4.45	111.2
Supplemented	3	2.5	10.3	257.5	4.88	122.0

TABLE 3. Changes in milk yields, dairy fat and milk protein content of buffalos control and experimental groups in an autumn series of experiments (daily mean data)

Buffalos Group	Number of animals	Yield of milk in l	Milk composition			
			Fat		Protein	
			In %	Total in g	In %	Total in g
At 10-fold reception of microelements						
Control	3	3.1	7.3	226.3	3.31	102.6
Supplemented	3	4.2	7.9	331.8	3.67	154.1
At 20-fold reception of microelements						
Control	3	3.1	7.9	244.9	3.87	120.0
Supplemented	3	5.3	8.0	424.0	3.94	208.8
At 30-fold reception of microelements						
Control	3	4.0	8.5	340.0	3.90	156.0
Supplemented	3	5.5	8.3	456.5	3.83	210.7

TABLE 4. Changes in milk yield, dairy fat and milk protein content in buffalos control and experimental groups of a winter series of experiments (daily mean data)

Buffalos Group	Number of animals	Yield of milk in l	Milk composition			
			Fat		Protein	
			In %	Total in g	In %	Total in g
At 10-fold reception of microelements						
Control	3	4.2	7.6	319.2	3.90	163.8
Skilled	3	5.6	7.5	425.6	4.03	225.7
At 20-fold reception of microelements						
Control	3	4.3	8.2	352.6	3.91	168.1
Skilled	3	5.7	8.0	456.0	3.78	215.5
At 30-fold reception of microelements						
Control	3	4.3	7.9	339.7	3.90	167.7
Skilled	3	5.4	8.0	432.0	3.87	209.0

Lactating buffalos receiving a mix of microelements in winter season, in comparison with animals from the control group did not change in respect to milk yield and fat content but the amount of milk protein was slightly reduced (Table 4).

In summarizing the results of our research and referring to previous studies we can tell with confidence, that top dressing buffalos and other agricultural animals receiving a mix of microelements consisting of chloride salts of cobalt, manganese and zinc, and also from potassium iodide positively affected the protein exchange and their efficiency. The positive influence of microelements was known to be effective only when the diet met all requirements showed by high-nutritive forages (Faye et al., 1992), i.e. contains necessary amount of various proteins, carbohydrates, lipids, various vitamins and mineral salts. Hygienic conditions of livestock maintenance and seasonal factors influencing the nutritive value of the diet had also important impact.

3.2. NEGATIVE ENVIRONMENTAL CONDITIONS

Some industrial enterprises cause negative influence on the environment of the western region of Azerbaijan. The greatest amount of industrial wastes came from an aluminium factory (Bachshalizade, 1998), (Bogdanovskiy, 1994). This area consisted of 250 ha of lands crossed by river Kura. After extracting the necessary components, waste products were accumulated in large areas around the plant (Figure 1). Thus, this flow negatively affected the ground, vegetation, water, air, the quality of agricultural products, and, hence, human health. Poisonous in structure and properties, the dust was carried up to 50–60 km. The chemical compound of waste products of the aluminium factory was mainly toxic (Tables 5 and 6). The degree of soil pollution was 38.64% and process of pollution continued (Photos 1 and 2).

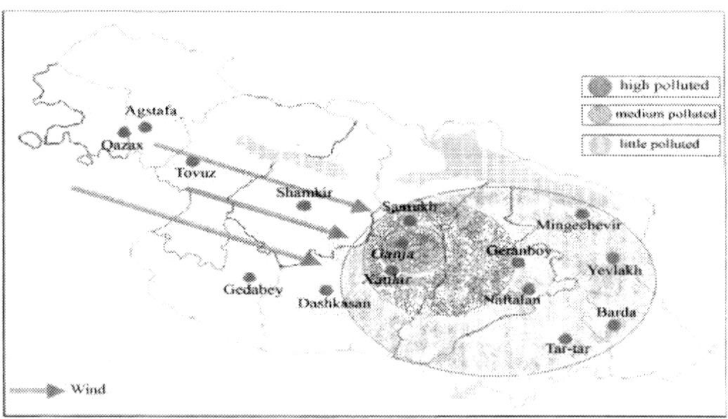

Figure 1. The general flow of environmental contamination from waste products of an aluminium factory

Photo 1. Polluted lands

Photo 2. Polluted lands

TABLE 5. Chemical compound of waste products (dry weight, %)

Components	Contents in %
SiO$_2$	75
Al$_2$O$_3$	14.85
Fe$_2$O$_3$	5.53
TiO$_2$	0.65
CaO	0.11
MgO	0.05
Mn	0.06
K$_2$O	0.74
Na$_2$O	0.22

TABLE 6. Content of waste products in heavy metals (dry weight, 10^{-3}%)

Components	Contents in %
Pd	0.30
Cd	0.052
Cu	6.3
Zn	4.4
Ni	1.9
Mo	4.34
Co	0.30
Cr	13

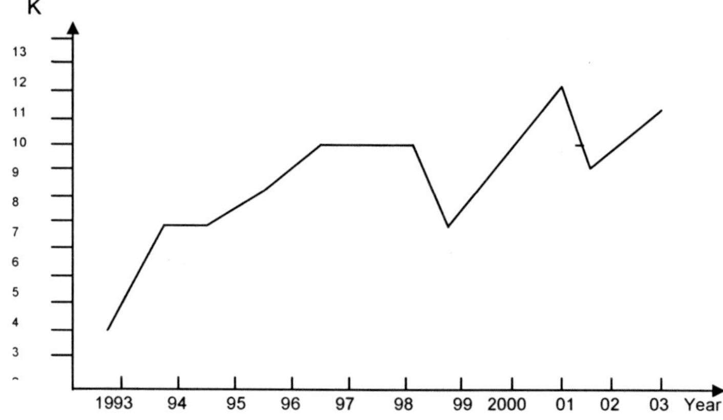

Figure 2. Change of the average concentration of phenols in water of the Kura River

The value of the Kura River for industry and agriculture in Azerbaijan was extremely important (Manafova, 2001). By virtue of its geographical position in our republic, the Kura River was also a source of potable water for more than 70% of its population.

Today more than 6 million cubic meters per day of the crude sewage was dumped into the Kura River and its main tributaries. More than 25,000 t a year of nitrogen-phosphoric compounds, hundreds tons of phenols (Figure 2), hydrocarbons, acids, alkalis, salts of heavy metals, also was running off into the Kura river (Table 7).

TABLE 7. Amount (mg/l) of some toxicants and hydro-chemical components in tributaries of the Kura River

The river	Phenols	Oil products	Copper	O_2	NO_2
Tovuzchay	0.027	0.1	0.02	6.0	0.07
Shamkirchay	0.018	0.02	0.018	8.2	0.1
Koshkarchay	0.014	0	0.017	8.0	0.2
Terter	0.015	0	0.01	7.4	0.03
Karkarchay	0.025	0.1	0.013	7.9	0.04
Turianchay	0.021	0.1	0.080	6.2	0.01
Karasu	0.013	0	0.007	6.8	0.01
Geokchay	0.011	0.05	0.01	7.0	0.02
Ganjachay	0.026	0.44	0.033	2.4	0.66
Damiraparanchay	0.013	0	0.010	7.1	0.01

3.3. THE AMOUNT OF MICROELEMENTS IN THE SOIL OF THE GANJA-KAZAKH ZONE OF AZERBAIJAN (MAMMADOV, 2000)

Manganese

In soils of the Ganja-Kazakh zone the amount of manganese was 400–1,650 mg/kg. The acquired form taken up by plants was 3–20 mg, with an average value of 12.8 mg. Plants of the Ganja-Kazakh zone were not deficient in manganese, but no supplementation occurred in livestock.

Zinc

The amount of zinc in soils of the Ganja-Kazakh zone of Azerbaijan was from 3.2 to 44.3 mg/kg. The amount acquired by plants was 0.4–5.2 mg, average value of 3.46 mg. The necessary amount of zinc in forages for animals was about 30 mg/kg.

Cobalt

Cobalt is the main element of cyanocobalamin (vitamin B12). The general amount of cobalt in the soils of the Ganja-Kazakh zone of Azerbaijan was 0.05–15 mg/kg. However some soils had little cobalt. The amount of cobalt here did not exceed 1.4 mg/kg. The average was 0.95 mg/kg and 0.12 mg was acquired by plants. he necessary amount of cobalt in animals forage was: cows – 20–40 mg, sheep – 10–20 mg/kg of forage.

Iodine

The amount of iodine in the soil of Azerbaijan was $15.1 \times 10^{-4}\%$. The high amount of iodine especially in the Caspian zone could be explained by the influence of Caspian Sea.

4. Conclusion

Supplementation of farm mammals and birds with a mix of some microelements renders positive influence on their dairy, meat, wool and other kinds of efficiency. On the basis of the study results we come to the conclusions that microelements could be managed properly to limit their actions for the benefit of animals by using them as supplement but the use of water and forage for animals in areas polluted by microelements must be clearly evaluated.

E.A. MAMMADOV AND I.J. ALVERDIYEV

References

Bogdanovskiy, G.A., 1994, Environmental Chemistry. Publishing house of Moscow State University, Moscow, Russia, 237 pp.

Bachshalizade, T.Z., 1998, Ecological value of waste influence on the soil-vegetable cover of adjoining territory in Ganja Aluminium factory. The abstract of the dissertation for a doctor's degree., Baku (Azerbaijan), 24 pp.

Faye, B., Saint-Martin, G., Cherrier, R., Ali Ruffa, M., 1992, The influence of high dietary protein, energy and mineral intake on deficient young camel: II. Changes in mineral status, *Comp. Biochem. Physiol.*, 102A: 417–424.

Mammadov, A.Y., 2000, Role of microelements in agriculture, Azerbaijan Technological University, Ganja, Azerbaijan.

Manafova, A., 2001, Importance of Exercising Regional Approach to the solution of water resources problems, Proc. of the Conference on Water Resources Management in the Countries of the South Caucasus, Tbilissi, Georgia, 6 July 2001, Legislation and Water, 10.

McDowell, L.R., 1992, Minerals in animal and human nutrition, Academic, San Diego, CA, 517 p.

URANIUM POLLUTION OF MEAT IN TIEN-SHAN

RUSTAM TUHVATSHIN, IGOR HADJAMBERDIEV, JURI
BIKHOVCHENKO

Toxic Action network of Central Asia, Bishkek, Kyrgyzstan

Abstract: Uranium in water, soil, fodder and food products (especially meat) was studied in areas of former Soviet
uranium industry in Tien-Shan 1950–1970. Uranium environment migration was very intensive in Tien-Shan, due to
peculiarities waters (upper and ground): hydro-carbon-calcium content, and low potassium medium. A content of
uranium in river-water from the areas was higher 30–100 times compared with rural regions of Russia. Lambs tissues
(wet weight) contain: 1.2 ± 0.15 mg/g uranium in Min-Kush, 0.06 ± 0.0002 mg/kg in Mailuu-Suu. Lambs skin, horn,
hoof contained 0.183 ± 0.007 mg/kg uranium. Meat of domestic animals was the only source of protein for local
people. It has been found that latent pre-illness was very wide spread in human population, but not registered in
official medical statistics. Health disorders (neutropenya, monocytosis, weak function of liver) have been shown by
blood laboratory tests. There are several unfortunate factors in Tien-Shan region: pollution, alpine hypoxia, lack of
iodine, protein and malnutrition. All these factors can lead together to low weak health of domestic animals and
human.

Keywords: Uranium, tails, meat of lamb, human health, Central Asia

1. Introduction

There were several pollution problems in Central Asia (CA).

1. Tank with dozens ton granulated cyanic products transport via deserts of
Kazakhstan and Uzbekistan to gold combines in Kyrgyzstan and Uzbekistan,
tremendous mining company -Kum-Tor Operating (now Centera) and incidents occurring in Issyk-Kul shore in 1998. The cyanide incidence and tremendous cyanide contain
wastage has been reviewed in several studies (Hadjamberdiev, 2000).

2. Other 300 warehouses of obsolete pesticides were still in non safety conditions in
each CA countries, and several illness and mortal incidences each year. From burial
places, there were outflows, and pesticides were exposed to atmospheric influence
(review of independent Kyrgyzstani "Azamat–Oil" corporation). It is known that in the
last years, certain measures were taken for storage of the pesticides forbidden for
application in the underground storehouses representing trenches and iron bunkers.
Despite of officially registered pesticides import reduction – there was a big unregistered
illegal obsolete pesticides imported from China, India and Pakistan.

3. Tien-Shan was a Uranium Boiler of USSR in 50th–60th.

There were 17 Stalin Price Laureates (scientists and engineers) for development of
uranium industry in the region. Dozens old uranium USSR warehouses still exist in
Tien-Shan and Pamir. Unfortunately, obsolete methods of uranium wastages keeping
had been used in 60th. Uranium wastage paste had been kept in the mountain gorges,
covered by concrete with sands and soil, with thin lie 6–10 m. There was special guard
service, which avoided animals and men (except miners) to close contact with tails
wastage in 60th–70th. But the guard system was destroyed at the beginning of 90th. We
didn't find protect service in the observed areas now (Photo 1). There are 48 tails
containing danger radioactive compounds in Tien-Shan. Total volume was 70 billion
cubic meters, sum activity 5500 Curie in Kyrgyzstan only. Several burst-accidents

occurred in Mailuu-Suu valley (in 1958 and in 1994) and in Ak-Tjuz region in North Kyrgyzstan (in 1964). There was uranium concentrating mill Kombinat (near Kara-Balta, Chui valley) which use ore from south Kazakhstan (Stepnogorsk mining's area).

In the present paper we reported our observations in two areas of uranium pollution for the last 2 years.

Photo 1. Former uranium plant near Mailuu-Suu River

2. Materials and Methods

For radiation detection, a radiometric equipment of former USSR was used: radiometer SRP-68-01 with counter BTGI-01 for gamma discrimination level 20 keV and diapason till 3,000 mRn/h. It has been using MSM from Speak with JMS-01-BM2, and GCG technique (Independent Lab Ilim under Acad Sci Kyrgyzstan and Organic Ins of Russian Acad Sci) for trace/radioactive elements analyses. We use mathematical comparison health-environment area and sub-areas (Lawson, 1999) and original eco-geographical classification of territory (Hadjamberdiev, 1996). Previous (latent) health-disorder manifestations have been estimated by original questionnaire had approved early (Hadjamberdiev, 2005a and b). Latest Manual International Chemist Analytic Association has been used for chemist analysis of human blood (glucose, cholesterol, liver enzymes ALT and AST).

3. Results and Discussion

Pulp was consistent of Mailuu-Suu tails. The pulp of tails N3, 7, 8, 18 contained 0.1–0.15% uranium (due obsolete technique of uranium elimination in years 50–60th). There were high content of other toxicants in the pulp (copper, cobalt, chromium, molybdenum, zinc). Our study showed that most important danger by uranium soil content was in area of tail number 3 of Mailuu-Suu region, where uranium was 35×10^{-6} g/g under surface one meter. The number was 35–50 times higher compared with geographical trend. Evidently, landslides and ground waters were spreading uranium and other above mentioned toxicants to land fields and to grass.

As it has been shown by our preliminary study uranium content in plants (dry residue) in Mailuu-Suu area was between very different, but sometimes very high. For example: in *Tacniatherum crinitum* ($2.29 \pm 0.03 \times 10^{-6}$) and *Atgilops triuncialis* (2.27×10^{-6}), but lower in *Cotoneaster suavis* ($0.28 \pm 0.001 \times 10^{-6}$ g/g).

Higher rate of accumulation was in grass with powerful root system. The fodder is a source for animal body building (cows and camels), so a source of pollution for meat and milk. It could lead to chronic disorders both in animals and human. Clear manifestation of disorders in local people was liver function disorder (Figs. 1–3).

Lambs meat contained 1.2 ± 0.15 mg/g uranium in Min-Kush, 0.06 ± 0.0002 mg/kg in Mailuu-Suu. Cow milk and meat in Min-Kush contained 2.27 ± 0.031 (P < 0.05) and 0.107 ± 0.001 (P < 0.05) mg/kg of wet weight. Lambs skin, horn, hoof in Mailuu-Suu contained 0.183 ± 0.007 (P < 0.05) mg/kg uranium. Meat of domestic animals was the only source of protein for local people. Human teeth uranium content (Mailuu-Suu): in milk-teeth $0.481 \pm 0.002 \times 10^{-6}$ g/g, in elderly people groups, from 0.7684×10^{-6} g/g to 0.6876×10^{-6} g/g (P < 0.05).

It has been observed several pre-illness and complete diseases in population from two studied old uranium areas (Min-Kush, Mailuu-Suu). It has been found what latent pre-illness widely spread in population, but such pre-illness status was not registered in official medical statistics.

The level of uranium (soil, grass, teeth) and inhabitants liver disorder and compliances (tachycardia, attacks of a migraine, insomnia) was supposed to be correlated. The laboratory tests shown: hyper-erythrocytemia, neutropenia, mono-cytosis, lymphocytosis, low level of whole blood protein. There was an evident low level of immunitary ability (lymphocytes, blood proteins, etc) in population of Min-Kush and Mailuu-Suu regions.

It was known that uranium content in upper water was higher 30–100 times compared with Central Russia (Lekarev, 1967). A high level of uranium migration in Tien-Shan environment was revealed. This migration was due to upper and ground water peculiarities: hydro-carbon-calcium content and low potassium medium.

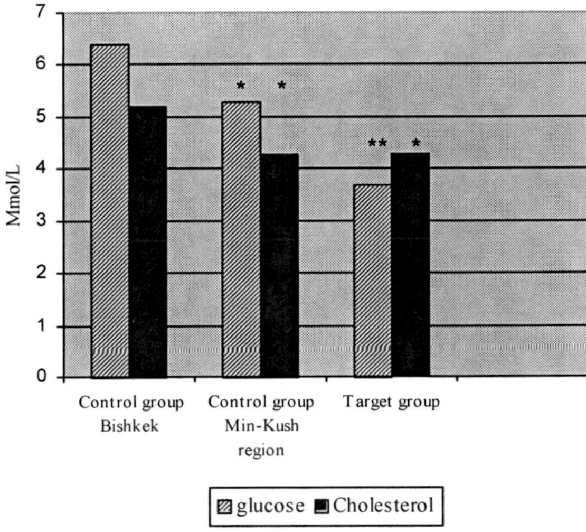

Figure 1. Tests of human lipid-carbohydrate liver functions in Min-Kush region *P < 0.05; **P < 0.01

But very few publications on animal and human health problems in former uranium industry areas were available except our recent reviews on problem in Central Asian toxic hot spots, including uranium (Hadjamberdiev, 1996; Hadjamberdiev and Tuhvatshin, 2002) based on official geological and medical statistics. Recent paper based on 3 years complex study of environment migration included impact on human body. On contrary, there was a lack of data concerning animal's contamination especially in meat consumed by humans where the uranium pollutant was concentrated.

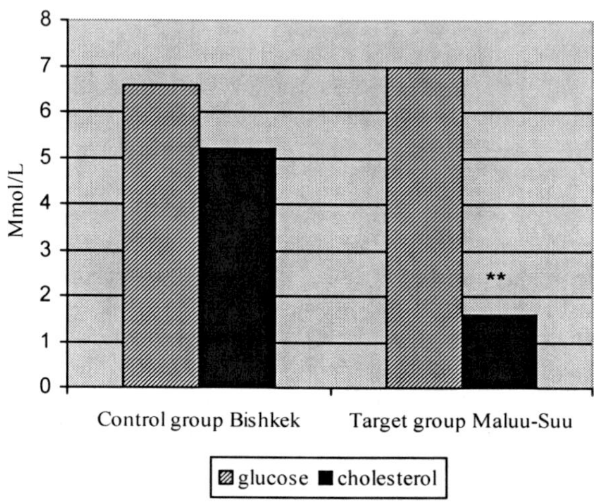

Figure 2. Tests of human liver lipid-carbohydrate function in Mailuu-Suu region. **P < 0.01

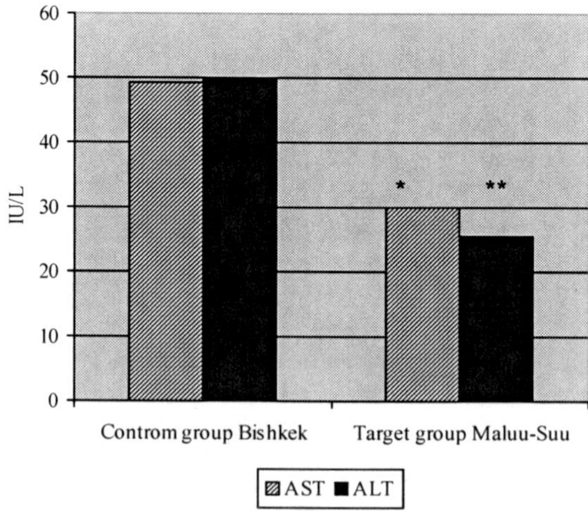

Figure 3. Enzyme balance (cell-internal and external) which marked human liver cells membrane damage in Mailuu-Suu region (AST: Aspartate Amino-transferarase; ALT: Alanine amino-transferase). *P < 0.05; **P < 0.01

4. Conclusion

A correlation between water contaminated by uranium (by underground infiltration from tails) on one hand and, and on the other hand the lamb and cow meat, cow milk, and human body and health status of human was probably existing, but subsequent epidemiological studies of this regions have to be achieved in these regions where uranium migration in ecosystems occurred.

Acknowledgments

This paper data base issued from our Project KR 766 of International Science Technical Centre (Bishkek).

References

Hadjamberdiev, I., 1996, Computer-cartography estimation on health-dangerous locus in Tien-Shen, In: Proc. 9th regional conf. IUAPPA, "Environment Impact Assessment", Prague, v. 4, pp. 703–705.

Hadjamberdiev, I., 2000, Cyanid in Tien-Shan — Danger. Bishkek, McArthurs funding (in Russian), 54 pp.

Hadjamberdiev, I., Tuhvatshin, R., 2002, Health disorder as consequence of environment disorder, in: Proc. Int conf. "Ecological Emergency in Kyrgyzstan", Bishkek: OSCE, pp. 104–106.

Hadjamberdiev, I., 2005a, Pesticides – Danger! Bishkek, 2005 (in Uzbek, Kyrgyz, Tadjik), 34 pp.

Hadjamberdiev, I., 2005b, Danger Chemical Substances in Central Asia. In: The Third World Congress on chemical, Biological and Radiological Terrorism, in Dubrovnik 17–23 September Aberdin (USA) Publ., ASA, 2006, pp. 123–127.

Lawson, A., 1999, Disease Mapping and Rick Assessment for Public Health, A. Lawson et al. (Eds). Wiley, Chichester, NY, 482 pp.

Lekarevn, V.S., 1967n Uranium biochemical chains in areas with different content of uraniumn Kandidat Dissertation, Moscow (in Russian), 112 pp.

SYNTHESIS AND RECOMMENDATIONS

BERNARD FAYE
Département environnement et société, CIRAD, Montpellier, France

The present book focused on the impact of pollution on animals and animal products included three main aspects:

1. Generalities on the role of institutions in the field of pollution assessment for animal and human health

The conferences involving this aspect were mainly presented at the opening session. After an introduction on the change in human health with the industrial revolution (T. Sharmanov), the structure of epidemiological control of animal health products in Kazakhstan was presented (A. Belonog) as well as the origin of the main contamination sources for the food products intended for human consumers and the consequences on their quality (M. Tultabayev). The strategy of the Kazakh government for improving the protection of human population face to important pollution occurring in the country was described (K. Ospanov), but it has been underlined the importance of the civil society and their public organizations to make lobbying in order to impose the realization of the international convention on pollutants (L. Astanina).

2. Characterization of the situation of environmental failures in Central Asia and western countries in relationship with human health and the role of animals as a risk for human health especially through the products (milk and meat)

This session aimed to have an overview of the contamination risk in different countries and to consider the tools and methods for a convenient estimation of the impact of pollution on animal and animal product as a step to human health.

Kazakhstan is an important oil producer and the impact of the oil pollution on the environment and the resources intended for animals is not well assessed in the country. The importance of the contamination of soil and water resources by oil wastes was enlightened (E. Ongarbaiev). This kind of pollution has a strong impact on food safety and it is obvious that a monitoring system must be set up, particularly because 70–100% of environmental contaminants may enter in human organism through the food intake (S. Sheveleva*). However, an important difficulty for a convenient monitoring is the quality of analytical procedure. So, it is crucial to develop reference analysis especially to detect molecules present in small quantities on food as dioxin and PCB (J.F. Focant). Only convenient equipments, good trained technicians and quality control all along the analytical procedures allow reliable data on the pesticides contents in food (J.F. Focant). It is obvious that this situation is not yet encountered in Central Asia. In Kazakhstan, for example the problems of control of biological activity of food supplements to nutrition are not solved and the lack of standard analytical method is one of the constraints to achieve such control (L. Kalamkarova). In Ukraine, this control overpass the only problem of analytical procedure, as the use and storage of prohibited and expired pesticides are common, provoking many troubles in animals, both pets and farm animals (T. Gurzhiy).

3. Relationships between soil contamination, plant contamination and animal products status

This session was based on several examples including different types of contamination (industrial and agricultural contamination by pesticides, heavy metals, radionuclides), but also contamination linked to feeding (feed supplementation). It is also considered, the relationships between the different compartment of a complex system going from soil and water (environment) to the animals and their products (meat and milk) intended to human consumption. In some cases, the impact on human health was considered.

In Morocco, it has been underlined the risk of pollution linked to mining activities as the exploitation of phosphates in Morocco provoking risk of fluorosis in animals like camels (M. Bengoumi). Heavy metals can also contaminated camels in desertic areas and it is possible to assess the mineral status of animals by analyzing the concentration of trace elements and heavy metals in blood although the lack of references in this species is especially remarkable (B. Faye). In order to assess heavy metals contamination on camel in Kazakhstan, different studies are currently achieved. Preliminary results are available giving data for copper, manganese, zinc, arsenic and lead in water and pastures (E. Diacono) as well as in camel milk and shubat (A. Meldebekova) with an analysis of the links between the different steps from the resources to the products. Here also, the lack of references is the main constraints to discuss the results. Moreover, it has been underlined the gap obtained by the results issued from different labs enlightening the necessity to standardize the analytical methods. The camel milk in Kazakhstan appears quite rich in minerals as calcium, phosphorus and iron. The main elements, calcium and phosphorus allow discriminating dromedary milk from Bactrian milk, more rich. However, the lead content is quite high (more than 0.2 ppm) in all the analyzed samples (G. Konuspayeva). The decontamination of non-conventional milks as camel or horse milk, currently consumed in Central Asia is all the more important because they are used as special foods for children (Y. Sinyavskiy*). The flow of contaminants from soil to food is in fact not easy to understand and the modeling (compartment model) is a convenient way to assess the risks from the contamination (C. Collins). To achieve such model, a lot of data are necessary, and the estimation of the flow between the compartment "soils" to the compartment "plant", then to the compartment "animal" has to be clarified. However, some transfer could be direct between soils to dairy ruminants as the direct soil intake can be quite important. The transfer to milk is highly variable. For PCBs, the transfer varies widely from 5% to 90% (S. Jurjanz). The contamination of animals is linked also to feed supplementation. In many case, the selenium supplementation is a quite importance in intensive systems (poultry, racing camels). For example, the assessment of selenium in final products is essential to control the impact of such supplementation on the selenium concentration in eggs (D. Cattaneo) and the risk linked to selenium toxicity by over intake in racing camel must be understood (R. Seboussi). Another risk, common in Central Asia, is the heavy metals contamination (cadmium, cobalt, manganese, lead, mercury) in nuclear sites. The contaminants can be present in milk and meat intended for human consumption with sometimes very high level above the normal standard (U. Kenesariev). The river could be also an important source of pollution for animal and human and the example of Danube River in Romania was well presented. However, interdisciplinary program can help for a decontamination of the sites all along the river (D. Galatchi*). In Italy, it was shown that the risk of contamination of cow milk by chemical pesticides was directly linked to the distance between the river and the contaminated pasture. The role of farming practices was underlined also (B. Ronchi). The detoxification is an important process to control food contamination. For example, control of aflatoxin contamination is possible with bioactive compound extract from *Silybum marianum* (silymarin), because the hepatoprotector activity (D. Tedesco). The impact of pollution on animal products was mainly approached through terrestrial mammals. Yet, water pollution is highly associated to fish and shellfish contamination. The organochlorine pollution was assessed in mussels reared in Crimea costal showing sometimes high concentrations of PCBs varying according the sites (L. Malakhova). The human diet is also a central point of the contamination risk in human. In Central Asia, the meat and meat product consumption is very high. Then, these products could be highly contaminated with N-nitroso compounds as the nitrates are widely used in agricultural process. The high consumption of such contaminants provokes risk of esophageal and gastric cancers (M. Aijanov*). The meat can be also contaminated by uranium, especially in areas where the water contains higher 30–100 times uranium compared with rural areas in Russia (I. Hadjamberdiev). This situation decreases the life expectancy in the region of nuclear site (N. Zhakashov). The relationships between trace elements in food products and human health have to be précised. Indeed, trace elements play an essential biological role but could be potentially toxic when their concentrations overpass limit values as it has been observed in Azerbaijan (E. Mammadov).

At the end of the presentations, two working groups were organized. The conclusion of the group 1 was reported by L. Malakhova from Ukraine and the second by B. Ronchi from Italia.

WG1. Three points have to be considered: (i) the sources of the contaminations, the content and distribution of the main pollutants (heavy metals, POPs, radioactive materials) in environment; (ii) the human as the final receptor of the food chain is human; (iii) between environment and human there are a lot of sections of the chain, by which the pollutants get to the human. One of the sections is the animals. Animals play the three main roles in that chain: (a) animals are the consumers of pollutants, and this is entailed by the multiple diseases and even deaths. This role is the subject of Veterinary Sciences, (b) animals are the food and food products suppliers for the human. This role of the animals is the subject of Nutritional and Food Sciences, and (c) another same important role of the animals is that they can serve as the markers of contamination level. With this role animals can be used to predict the environmental changes and to avoid big environmental crisis.

Then the strategy must be constructed with those three roles: (i) we should launch new and continue existing studies of the animal diseases and their prevention, (ii) we should pay more attention to Food Safety, (iii) we should evaluate the animal' role in pollutants transformation from the sources of contamination to the human. In conclusion, the recommandations of this working group were:

1. To develop regional program of the pollutants monitoring in animal products
2. To identify the valid regional markers and indicators of contamination
3. To develop the common methodology for contamination assessment in environment and animals and to evaluate the population health status in region
4. To develop the recommendations for diminishing of contamination risk of animal products
5. To give information on the situation with the food products contamination and possible risk diminishing ways to the government bodies and non-public organizations

6.   To develop the project funded by international organizations within the frames of the proposals

WG2. To reinforce research on the interactions between Livestock Production Sciences and pollution, it is necessary to propose four topics: (i) for implementing exchanges between researchers from different countries in order to compare data and information, to optimize research efficiency, to transfer techniques and to develop multidisciplinary approaches, (ii) for improving large scale studies using information systems by the development of model of risk assessment, the collection of data on pollution sources and impact, the characterization of polluted areas, the identification of the contaminants concentration in foods and by the studies on the relation between agricultural practices and pollution, (iii) for improving analytical methods (quality standard), it is necessary to validate used methods, to improve the existing methods, to diminish their cost and rapidity, to identify and use biomarkers of exposure and biologicals, especially domestic animals, (iv) for assessing the interaction between animals and pollution.

To achieve these topics, it is recommended to act in five steps:
1.   To identify the pollution problem in a specific area or different zones
2.   To characterize the problem of the pollution by collecting reliable data
3.   To identify a multidisciplinary team (within or between countries)
4.   To report the findings
5.   To make recommendations for policy makers and NGOs.

Finally, the main recommendation was to try an answer for the future call of EU projects in 2008 in the frame of the 7th Program. The first point is to identify a coordinator, to establish the procedures and steps, to make proposals for each partner and to develop the capacity building of the partners. A first concept note must be proposed by the directors of the present workshop.

*Those conferences were not presented in the present book as they did not reach the editor's office in time.

List of speakers

Name	Country	Full official address
Co-directors		
Faye Bernard faye@cirad.fr	France	UR18 – CIRAD – TA C-DIR/B Campus international de Baillarguet 34398 Montpellier cedex 5 France
Synyavskiy Yuri Sinyavskiy@list.ru	KZ	Kazakh Academy of Nutrition 050008, Almaty, Klochkov st. 66
Speakers and Lecturers		
Aijanov M.M. Aidjanov@yahoo.com	KZ	Kazakh Academy of Nutrition 050008, Almaty, Klochkov st. 6
M.K. Amrin	KZ	Kazakh National Medical University, Hygiene and Ecology Chair Tole bi 88, 050060 Almaty
Astanina Lidiya greenwomen@nursat.kz	KZ	Lydia Astanina Head of the Greenwomen Ecological News Agency Koktem 2, 2-73, Almaty, 480070 KAZAKHSTAN
Belonog Anatoly a.belonog@mz.gov.kz	KZ	Comity of State sanitary-epidemiological control of Ministry of Healthcare of Kazakhstan Astana 010000, 35th st. 2/5
Ivashchenko Anatoly a_ivashchenko@mail.ru	KZ	Kazakh National University of Al Farabi 050078 Almaty, 71 Al Farabi av
Kalamkarova Lucia olga_bagryanseva@mail.ru	KZ	Kazakh Academy of Nutrition 050008, Almaty, Klochkov st. 66
Kenesariev U.I.	KZ	Kazakh National Medical University, Hygiene and Ecology Chair Tole bi 88, 050060 Almaty
Konuspayeva Gaukhar konuspayevags@hotmail.fr	KZ	Kazakh National University of Al Farabi 050078 Almaty, 71 Al Farabi av
Meldebekova Aliya alisha_kz@mail.ru	KZ	Kazakh National University of Al Farabi 050078 Almaty, 71 Al Farabi av
Ongarbaev Yerdos Yerdos_ong@kazsu.kz	KZ	Kazakh National University of Al Farabi 050078 Almaty, 71 Al Farabi av
Ospanov K.S. rses@netmail.kz	KZ	Republic Epidemiological Station Auezov St. 84, Almaty 050008
Sharmanov Torekeldi Sharmanov.t@mail.ru	KZ	Kazakh Academy of Nutrition 050008, Almaty, Klochkov st. 66
Tultabaev Muhtar	KZ	Ministry of Ecology and Environment
Mammadov Elshad eco@azeurotel.com	Azerbaijan	Azerbaijan Technological University 103, H. Aliyev Ave., AZ2011 Ganja, Azerbaijan
Focant Jean-Francois JF.Focant@ulg.ac.be	Belgium	Mass spectrometry lab, Biological & organic analytical chemistry, University of Liege allée de la Chimie, B6c, B-4000 Liège (Sart-Tilman), Belgium
Seboussi Rabiya[a] rseboussi@yahoo.com	Canada	Saint-Hyacinthe veterinary school, Montreal University
Diacono Emilie emiliediacono@no-log.org	France	UR18 – CIRAD – TA C-DIR/B Campus international de Baillarguet 34398 Montpellier cedex 5 France

Jurjanz Stephan stefan.jurjanz@ensaia.inpl-nancy.fr	France	USC INRA/INPL/ENSAIA/Univ. Nancy 1: Animal et Fonctionnalités des Produits Animaux ENSAIA 2 avenue de la Forêt de Haye – BP 17254505 Vandoeuvre-Les-Nancy Cedex, France
Cattaneo Donatta donata.cattaneo@unimi.it	Italy	VSA-Dept. of Vet. Sci. and Technol. for Food Safety, University of Milan, Via Trentacoste 2-20134, Milan, Italy
Ronchi Bruno ronchi@unitus.it	Italy	Department of animal production, Tuscia University Via De Lellis s.n.c, 01100 Viterbo, Italy
Tedesco Doriana doriana.tedesco@unimi.it	Italy	Department of Sciences and technologies, University of Milan Via trentacoste 2 20134 Milan, Italy
Al-Alawi Mutaz[b] alawi1979@yahoo.com	Jordan	Environment Sciences and management P.O. Box 16 Karak 61110 Jordan
Hadjamberdiev Igor igorho@mail.ru	Kyrgyzstan	Head of laboratory of Ecological monitoring of Goi-gol lake Chairman of Ecological Found Toxic Action network of Central Asia, 1451, Bishkek 720040, Kyrgyzstan
Bengoumi Mohammed m.bengoumi@iav.ac.ma	Morocco	Département de Pharmacie, Toxicologie et Biochimie, Institut Agronomique et Vétérinaire Hassan II BP 6202 Rabat-Instituts, 10100 Rabat, Maroc.
Galatchi Liviu-Daniel galatchi@univ-ovidius.ro	Romania	Ovidius University of Constanta, Romania
Sheveleva S.A. hotimchenko@ion.ru	Russia	Head of Laboratory of Nutrition Toxicology, Russian Institute of Nutrition, Russian Academy of Medical Science Moscow, Ustinsky proezd 2/14
Collins Cris c.d.collins@reading.ac.uk	UK	Reader in Soil ScienceDepartment of Soil Science, School of Human and Environmental Sciences, The University of Reading, Whiteknights, Reading, RG6 6DW, UK
Gurzhiy Tamara tgurzhiy@gmail.com	Ukraine	Independent Agency for Ecological Information, Kharkiv, Ukraine
Malakhova Ludmila malakh2003@list.ru	Ukraine	Dept. of Radiation and Chemical Biology, Institute of Biology of the Southern Seas, National Academy of Sciences of Ukraine

[a]Not present but her PowerPoint was presented by B. Faye, co-author
[b]Not present, but has sent his PowerPoint presentation and his full paper

List of other participants

Name	Country	Full official address
Absemetova M.A. rses@netmail.kz	KZ	Republic Epidemiological Station Auezov St. 84, Almaty 050008
Akhatova Z.M. nadira_musaeva@rambler.ru	KZ	Livestock and Veterinary Medicine Insitute Jandosova 51 480035 Almaty
Akhmetsadykov Nurlan info@kgau.almaty.kz	KZ	Kazakh Agrarian University Abai 8, 480008 Almaty
Baubekova Almagul baubekova@kazsu.kz	KZ	Kazakh National University of Al Farabi 050078 Almaty, 71 Al Farabi av
Izmailova Natalia n.izmailova@atom.almaty.kz	KZ	Kazakhstan Atomic Energy Committee Lisy Chaikinoy St. 4, 050020 Almaty
Mansurov Zulkhair mansurov@kazsu.kz	KZ	Kazakh National University of Al Farabi 050078 Almaty, 71 Al Farabi av
Moussaev Z.M. nadira_musaeva@rambler.ru	KZ	Livestock and Veterinary Medicine Insitute Jandosova 51 480035 Almaty
Namietov Askar info@kgau.almaty.kz	KZ	Kazakh Agrarian University Abai 8, 480008 Almaty
Narmuratova Miramkul miramkul@rambler.ru	KZ	Kazakh National University of Al Farabi 050078 Almaty, 71 Al Farabi av
Serikbaeva Asia mamula@ok.kz	KZ	Kazakh Agrarian University Abai 8, 480008 Almaty
Souleymenova Zh. Souleymenova.z@mail.ru	KZ	Kazakh Academy of Nutrition 050008, Almaty, Klochkov st. 66
Tsoi Igor tsoi.i@mail.ru	KZ	Kazakh Academy of Nutrition 050008, Almaty, Klochkov st. 66